BIOMETHANE

Developments and Prospects

BIOMETHANE

Developments and Prospects

Edited by

Dr. Sonil Nanda

Dr. Prakash K. Sarangi

APPLE ACADEMIC PRESS

First edition published 2022

Apple Academic Press Inc.
1265 Goldenrod Circle, NE,
Palm Bay, FL 32905 USA

4164 Lakeshore Road, Burlington,
ON, L7L 1A4 Canada

CRC Press
6000 Broken Sound Parkway NW,
Suite 300, Boca Raton, FL 33487-2742 USA

2 Park Square, Milton Park,
Abingdon, Oxon, OX14 4RN UK

Library and Archives Canada Cataloguing in Publication

Title: Biomethane : developments and prospects / edited by Dr. Sonil Nanda, Dr. Prakash K. Sarangi.

Names: Nanda, Sonil, editor. | Sarangi, Prakash Kumar, editor.

Description: First edition. | Includes bibliographical references and index.

Identifiers: Canadiana (print) 2021038980X | Canadiana (ebook) 20210389850 | ISBN 9781774639825 (hardcover) | ISBN 9781774639832 (softcover) | ISBN 9781003277163 (ebook)

Subjects: LCSH: Methane—Biotechnology. | LCSH: Methane as fuel. | LCSH: Biomass energy.

Classification: LCC TP761.M4 B56 2022 | DDC 665.7/76—dc23

Library of Congress Cataloging-in-Publication Data

CIP data on file with US Library of Congress

ISBN: 978-1-77463-982-5 (hbk)
ISBN: 978-1-77463-983-2 (pbk)
ISBN: 978-1-00327-716-3 (ebk)

About the Editors

Sonil Nanda, PhD
Research Associate, Department of Chemical and Biological Engineering, University of Saskatchewan, Saskatoon, Saskatchewan, Canada

Dr. Sonil Nanda is a Research Associate in the Department of Chemical and Biological Engineering at the University of Saskatchewan, Saskatoon, Saskatchewan, Canada. He has published over 120 peer-reviewed journal articles, 70 book chapters and has presented at many international conferences. His research areas are related to the production of advanced biofuels and biochemical through thermochemical and biochemical conversion technologies such as gasification, pyrolysis, carbonization, torrefaction, and fermentation. He has gained expertise in hydrothermal gasification of various organic wastes and biomass, including agricultural and forestry residues, industrial effluents, municipal solid wastes, cattle manure, sewage sludge, food wastes, waste tires, and petroleum residues to produce hydrogen fuel. His similar interests are also in the generation of hydrothermal flames for the treatment of hazardous wastes, agronomic applications of biochar, phytoremediation of heavy metal contaminated soils, as well as carbon capture and sequestration.

Dr. Nanda is the editor of books entitled *New Dimensions in Production and Utilization of Hydrogen* (Elsevier), *Recent Advancements in Biofuels and Bioenergy Utilization* (Springer Nature), *Biorefinery of Alternative Resources: Targeting Green Fuels and Platform Chemicals* (Springer Nature), *Fuel Processing and Energy Utilization* (CRC Press), *Bioprocessing of Biofuels* (CRC Press) and *Biotechnology for Sustainable Energy and Products* (I.K. International Publishing House Pvt. Ltd.). Dr. Nanda is a Fellow Member of the *Society for Applied Biotechnology* in India and a Life Member of the *Indian Institute of Chemical Engineers*, *Association of Microbiologists of India*, *Indian Science Congress Association*, and the *Biotech Research Society of India*. He is also an active member of several chemical engineering societies across North America, such as the *American Institute of Chemical Engineers*, the *Chemical*

Institute of Canada, the *Combustion Institute-Canadian Section*, and *Engineers Without Borders Canada*. Dr. Nanda is Assistant Subject Editor of the *International Journal of Hydrogen Energy* (Elsevier) as well as an Associate Editor of *Environmental Chemistry Letters* (Springer Nature) and *Applied Nanoscience* (Springer Nature). He has also edited several special issues in renowned journals such as the *International Journal of Hydrogen Energy* (Elsevier), *Chemical Engineering Science* (Elsevier), *Biomass Conversion and Biorefinery* (Springer Nature), *Waste, and Biomass Valorization* (Springer Nature), *Topics in Catalysis* (Springer Nature), *SN Applied Sciences* (Springer Nature), and *Chemical Engineering and Technology* (Wiley).

Dr. Nanda received his PhD degree in Biology from York University, Canada; MSc degree in Applied Microbiology from Vellore Institute of Technology (VIT University), India; and BSc degree in Microbiology from Orissa University of Agriculture and Technology, India. He has worked as a Postdoctoral Research Fellow at York University, the University of Western Ontario and the University of Saskatchewan in Canada.

Prakash K. Sarangi, PhD
Scientist, Central Agricultural University, Imphal, Manipur, India

Dr. Prakash K. Sarangi is a Scientist with a specialization in Food Microbiology at the Central Agricultural University in Imphal, Manipur, India. His current research is focused on bioprocess engineering, renewable energy, biofuels, biochemicals, biomaterials, fermentation technology, and postharvest engineering and technology. He has more than 10 years of teaching and research experience in biochemical engineering, microbial biotechnology, downstream processing, food microbiology, and molecular biology.

Dr. Sarangi has served as a reviewer for many international journals and has authored more than 70 peer-reviewed research articles and 50 book chapters. Dr. Sarangi has edited the following books entitled *Recent Advancements in Biofuels and Bioenergy Utilization* (Springer Nature), *Biorefinery of Alternative Resources: Targeting Green Fuels and Platform Chemicals* (Springer Nature), *Fuel Processing and Energy Utilization* (CRC Press), *Bioprocessing of Biofuels* (CRC Press), and *Biotechnology for Sustainable Energy and Products* (I.K. International Publishing House Pvt. Ltd.). Dr. Sarangi serves as an academic editor for PLOS One journal. He is associated with many scientific societies as a fellow member (Society for Applied Biotechnology) and Life Member (Biotech Research Society of India; Society for Biotechnologists of India; Association of Microbiologists of India; Orissa Botanical Society; Medicinal and Aromatic Plants Association of India; Indian Science Congress Association; Forum of Scientists, Engineers, and Technologists; International Association of Academicians and Researchers; Hong Kong Chemical, Biological, and Environmental Engineering Society; International Association of Engineers; and Science and Engineering Institute).

Dr. Sarangi received his PhD degree in Microbial Biotechnology from the Department of Botany at Ravenshaw University, Cuttack, India; MTech degree in Applied Botany from the Indian Institute of Technology Kharagpur, India; and MSc degree in Botany from Ravenshaw University, Cuttack, India.

Contents

Contributors

Ajay K. Dalai
Department of Chemical and Biological Engineering, University of Saskatchewan, Saskatoon, Saskatchewan, Canada

Shankar Swarup Das
Department of Farm Machinery and Power Engineering, Central Agricultural University, Ranipool, Gangtok, India

Ajinath Dukare
Chemical and Biochemical Processing Division, Indian Council of Agricultural Research - Central Institute for Research on Cotton Technology, Mumbai, Maharashtra, India

Rajesh Kanna Gopal
Department of Plant Biology and Plant Biotechnology, Presidency College, Chennai, Tamil Nadu, India

Nupur Kesharwani
Department of Civil Engineering, National Institute of Technology, Raipur, Chhattisgarh, India

Janusz A. Kozinski
Faculty of Engineering, Lakehead University, Thunder Bay, Ontario, Canada

Roshan Kumar
Department of Human Genetics and Molecular Medicine, Central University of Punjab, Bathinda, Punjab, India

Rudra Narayan Mohapatro
Department of Chemical Engineering, C.V. Raman Global University, Bhubaneswar, Odisha, India

Alivia Mukherjee
Department of Chemical and Biological Engineering, University of Saskatchewan, Saskatoon, Saskatchewan, Canada

Sonil Nanda
Department of Chemical and Biological Engineering, University of Saskatchewan, Saskatoon, Saskatchewan, Canada

Jude A. Okolie
Department of Chemical and Biological Engineering, University of Saskatchewan, Saskatoon, Saskatchewan, Canada

Nidhi Pareek
Department of Microbiology, Central University of Rajasthan, Ajmer, Rajasthan, India

Kunwar Paritosh
Center for Energy and Environment, Malaviya National Institute of Technology, Jaipur, Rajasthan, India

Biswa R. Patra
Department of Chemical and Biological Engineering, University of Saskatchewan, Saskatoon, Saskatchewan, Canada

Falguni Pattnaik
Center for Rural Development and Technology, Indian Institute of Technology Delhi, New Delhi, India

Preethy P. Raj
Department of Biotechnology, University of Madras, Chennai, Tamil Nadu, India

Prakash K. Sarangi
Directorate of Research, Central Agricultural University, Imphal, Manipur, India

Ranjita Swain
Department of Chemical Engineering, C.V. Raman Global University, Bhubaneswar, Odisha, India

Apoorva Upadhyay
Center for Energy and Environment, Malaviya National Institute of Technology, Jaipur, Rajasthan, India

Vivekanand Vivekanand
Center for Energy and Environment, Malaviya National Institute of Technology, Jaipur, Rajasthan, India

Abbreviations

AB	anaerobic bioreactor
AD	anaerobic digestion
AF	anaerobic filter
BEP	break-even point
BioSNG	biosynthetic natural gas
BMP	biochemical methane potential
BOD	biological oxygen demand
BPGTP	biogas-based power generation (off-grid) and thermal energy application program
C/N	carbon-to-nitrogen
CAR	anaerobic compartmentalized reactor
CNG	compressed natural gas
CoB	coenzyme B
COD	chemical oxygen demand
COS	carbonyl sulfide
CS	corn stalk
CSTR	continuous flow stirred tank reactor
DCF	discounted cash flow
DEPG	dimethyl ethers of polyethelyne glycol
DGA	diglycolamine
DME	dimethyl ether
EGSB	extended granular sludge bed
FT	Fischer-Tropsch Synthesis
GHG	greenhouse gas
GLS	gas-liquid-solid
H_2S	hydrogen sulfide
HRT	hydraulic retention time
IC	internal circulation
IEA	International Energy Agency
IRENA	International Renewable Energy Agency
LPG	liquefied petroleum gas
MEC	microbial electrolysis cell
NBMMP	national biogas and manure management program
NNBOMP	new national biogas and organic manure program

NPV	net present value
OFMSW	organic fraction of municipal solid waste
PEG	polyethylene glycol
PM	particulate matter
PVC	polyvinyl chloride
RS	rice straw
SNG	synthetic natural gas
SPAC	spiral automatic circulation
SRT	solids retention time
SS	suspended solid
STP	sewage treatment plant
SUFR	spiral up-flow reactor
TEA	techno-economic analysis
TPC	total product cost
TS	total solid
UASB	up-flow anaerobic sludge blanket
USSB	up-flow staged sludge bed
VBP	volumetric biogas production
VFA	volatile fatty acid
WS	wheat stalk

Preface

Since the Industrial Revolution, the heat and power industries, the automotive sector, commercial manufacturing sector, and the energy industry have heavily relied on fossil fuels and their derivatives. Meanwhile, the environmental concerns and adverse conditions—such as global warming, climate change; pollution of air, water, and soil; rise in atmospheric temperature; alterations in weather patterns; seasonal variations; acid rain; ozone lay depletion as well as the catastrophic shifts in the natural ecosystems and habitats—are also experiential across different geographies of the world. Although these negative environmental conditions are interrelated, both direct and indirect correlations could be ascertained to the exploiting usage of fossil fuels. The extraction, processing, and burning of fossil fuels generate massive amounts of greenhouse gas emissions, which are the leading cause of global warming and most of the other aforementioned environmental concerns. The health hazards to humans and animals caused by atmospheric, terrestrial, and aquatic pollutions are also ascribed to anthropogenic activities and greenhouse gas emissions. Therefore, it is high time to seek alternative energy sources that can gradually replace fossil fuels to reduce environmental impacts and strategize to mitigate global warming and climate change. This book covers some recent advances in the production and utilization of biomethane.

Chapter 1 by Das et al. broadly focuses on biogas (or biomethane), emphasizing the different biomass utilization for industrial and domestic applications. Chapter 2 by Upadhyay et al. describes the characteristics, parameters, and process design of anaerobic digesters for biomethane production from waste biomass. Chapter 3 by Gopal et al. deals with advanced genetic engineering tools and techniques implemented to enhance biomethanation and biomethane production. Chapter 4 by Mukherjee et al. makes a state-of-the-art review of anaerobic digestion of biogenic solid wastes, the impact of different chemical pretreatment processes and products, and the influence of operating parameters on biomethane yields. Chapter 5 by Patra et al. differentiates between the thermochemical technologies (e.g., gasification and pyrolysis) and biological technologies (e.g., anaerobic digestion) for biomethane production. The chapter also assesses some recent advancements in biomethane production along with its socio-economic impacts and applications. Chapter 6 by Paritosh et al. discusses gasification technology and syngas cleaning for biosynthetic natural gas production. The chapter also sheds light on the use of catalysts for enhanced synthetic natural gas production. Chapter 7 by Sarangi and Nanda provides an overview of biohythane fuel produced from microbial fermentative pathways. Chapter 8 by Swain et al. describes the socio-economic and techno-economic impacts of biohydrogen and biomethane production technologies.

This book provides a comprehensive synopsis on the topics mentioned above for applications in several cross-disciplinary areas of biotechnology, fermentation technology, bioprocess engineering, chemical engineering, and environmental technology with a common interest in biofuels and bioenergy.

The editors are grateful to all the authors for contributing their quality scholarly materials to develop this book. We express our sincere thanks to Ashish Kumar and Sandy Jones Sickels from Apple Academic Press, Inc. for their enthusiastic assistance while developing this book.

Dr. Sonil Nanda
Research Associate
Department of Chemical and Biological Engineering
University of Saskatchewan
Saskatoon, Saskatchewan, Canada
E-mail: sonil.nanda@outlook.com
ORCID: https://orcid.org/0000-0001-6047-0846

Dr. Prakash K. Sarangi
Scientist, Directorate of Research
Central Agricultural University
Imphal, Manipur, India
E-mail: sarangi77@yahoo.co.in
ORCID: https://orcid.org/0000-0003-2189-8828

CHAPTER 1

Utilization of Waste Biomass Resources for Biogas Production

SHANKAR SWARUP DAS,[1] PRAKASH K. SARANGI,[2] and SONIL NANDA[3]

[1]*Department of Farm Machinery and Power Engineering, Central Agricultural University, Ranipool, Gangtok, India*
E-mail: shankarswarup@gmail.com (Shankar Swarup Das)

[2]*Directorate of Research, Central Agricultural University, Imphal, Manipur, India*

[3]*Department of Chemical and Biological Engineering, University of Saskatchewan, Saskatoon, Saskatchewan, Canada*

ABSTRACT

Alternative fuel generation has drawn the attention of researchers due to the forthcoming warning of fossil fuel insufficiency. This has led to the utilization of various alternative and sustainable biomasses for biofuel production. Biofuel derived from biomass are considered the major solution to meet the future challenges raised by fossil fuel, unusual climate change, and greenhouse gas emission. Hence, biomass, which is rich in carbohydrates, can also be effectively converted into hydrocarbons. Biomass can be transformed into biofuels such as bioethanol, biodiesel, biohydrogen, and biogas to use it widely for transportation, industrial, and domestic applications. Many developments in the production of biofuels and biogas have been reported in recent years. This chapter broadly focuses on the production of biogas with emphasis on the different biomass utilization for various industrial and domestic applications.

1.1 INTRODUCTION

Fossil fuel contributes a major part in the scenario of energy supply in the modern world in the form of coal, crude oil and gasses (Sarangi et al., 2012; Nanda et al., 2015). The demand for energy supply has been extensively increased due to the inherent power generation, industrialization, and transportation. Coal as solid fuel is mainly used for power generation in the thermal power plants, whereas crude oil in the form of petrol and diesel are used for the transportation sector and the gasses for domestic purpose as cooking gas (Sarangi and Sahoo, 2010; Rana et al., 2020). The escalating usage of fossil fuel and natural gas has led to the environmental problems of greenhouse gas emission and insufficiency from the sources. Many researchers have reported that fossil fuel sources throughout the world will last up to 25 years, and there is a need for alternative sources such as renewable energy (Sarangi and Nanda, 2018, Sarangi and Nanda, 2019; Bhatia et al., 2020).

Biofuels are the most prominent source of renewable energy and are expected to be the major replacement of the current fossil fuel. These fuels are gaining more popularity due to their environment-friendly and sustainable nature (Nanda et al., 2018; Nanda et al., 2019). It is also observed that the biomass takes an equivalent amount of CO_2 during its growth and release in consumption, making biofuels carbon-neutral (Baçaoui et al., 1998). These are generally considered as organic such as forest residues, agricultural residues, grass cuttings, animal manure, sewage sludge, etc., (Gong et al., 2017; Nanda et al., 2016; Nanda et al., 2020; Sarangi et al., 2020; Sarangi and Nanda, 2020; Siang et al., 2020).

In this chapter, the overall possible paths for the production of biogas with the technologies used for the conversion of biomass to biofuel is discussed. Historical indication shows that the biogas was mainly used for heating water for bath purposes during the mid of the sixteenth century at Persia (Cheng, 2010). Biogas predominantly contains CH_4, which is generally produced by the anaerobic digestion process where the organic wastes are degraded by methanogenic bacteria without oxygen (Bessou et al., 2009). It is considered clean and renewable energy that which is a substitute for natural gas to be used for cooking, water heating and electricity generation (Andrea and Fernando, 2012).

Biogas is available in the gaseous form at room temperature and pressure, unlike the liquid form of liquefied petroleum gas (LPG). Anaerobic digestion or methanation is performed in four basic steps such as hydrolysis, acetogenesis, acidogenesis, and methanogenesis. In the case of the hydrolysis process, the rate of liberation of gas is limited because the polymers contain complex polymers. Biomass can be subjected to microbial attack in case of the pretreatment processes (Hall and Scrase, 1998). Hence the pretreatment can be physically done like irradiation, biological treatments by enzymes or fungus, chemical treatment with oxidation, acids, alkalis. Sometimes it can be done by a combination of all these processes (Omer, 2012). Acidogenes in the anaerobic digestion process is nothing but the acidogenic microorganisms further spitted into biomass after hydrolysis. In this case, the fermentative bacteria produce NH_3, H_2, CO_2, H_2S, fatty acids, alcohols and carbonic acids apart from producing an acidic environment. Mostly the acidogenic bacteria break down the organic matter of the biomass to produce a large quantity of useful methane under the acetogenesis process. The fundamental steps of the conversion of biomass to biogas using a biodigester or anaerobic digester are shown in Figure 1.1.

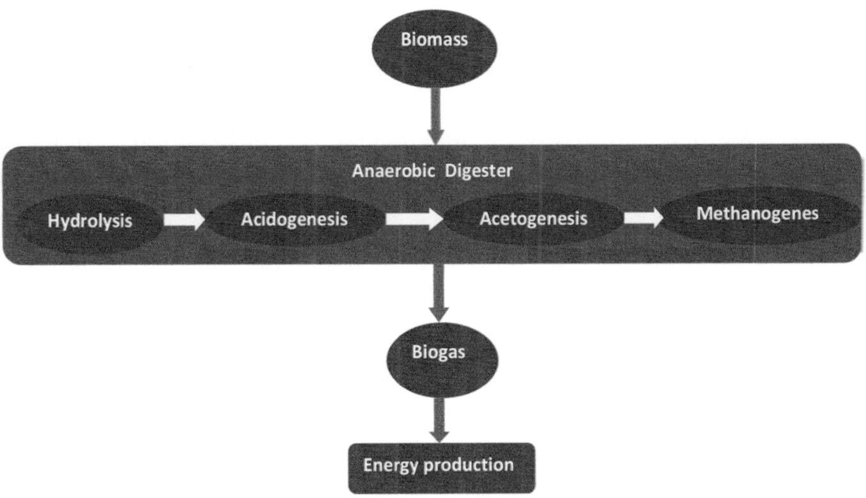

FIGURE 1.1 Fundamental steps of biomethanation in a biodigester.

The final step of the anaerobic digestion process is methanogenesis, which constitutes the final products of acetogenesis, and a few intermediate products of hydrolysis and acidogenesis (Omer, 2015). The main mechanism in methanogenesis is to convert CO_2 into methane and by consuming

H_2, which involves the path of acetic acid, which gives two products in anaerobic digestion like CH_4 and CO_2 (Taherzadeh et al., 2008).

1.2 BIOCONVERSION PROCESS

The transformation and production of biomass into useful biofuel products needs deep knowledge in chemistry, engineering, and control system. The types of bio-refineries based upon varieties of the raw materials, technologies, products, and processes are classified as first, second, and third-generation refineries (Rasslavicius et al., 2011). Some of the common technologies used for the conversion of biomass into biofuel and biogas are biological, physical, chemical, and thermal, depending on the type of product.

The biological conversion can be classified into anaerobic digestion, saccharification, and dark/photo fermentation, which tends to produce biomethane, ethanol, and biohydrogen, respectively (Nanda et al., 2014). Furthermore, the physical conversion can be classified into mechanical extraction, briquetting, and distillation. Bioconversion requires pretreatment and hydrolysis, which tends to release monomeric cellulose and hemicellulose for microbial fermentation. The thermochemical conversion can be classified into pyrolysis, liquefaction, and gasification to produce bio-oil, bio-crude oil and syngas, respectively (Okolie et al., 2021; Parakh et al., 2020).

1.3 BIOGAS PRODUCTION

Anaerobic digestion has traditionally aided low and middle-income countries, particularly the rural economies to sustainably manage the biogenic wastes, generate revenues and local employment while producing clean fuels to meet the domestic energy demands. The utilization of biogas is similar to the natural gas commonly used for cooking, heating or as a gaseous fuel for vehicles. Mainly, it contains CH_4, CO_2, water vapor and traces of N_2, NH_3, H_2, and H_2S. The energy content of biogas mainly depends upon the quantity of CH_4 present in it (Omer and Yemen, 2003). Hence, a high amount of CH_4 is always desirable. Care must be taken to avoid the water content and CO_2, and also to minimize the sulfur content for the engines of vehicles considering the pollution stage norms. The biogas has an average calorific value of 21–25 MJ/m^3, while 1 m^3 of biogas is equivalent to 0.5 L of diesel fuel with 6 kWh (FNR, 2009).

The biogas production rate of a plant depends upon the design, feedstock, temperature, and holding time (Martin et al., 2019). For example, the common feedstocks used in anaerobic digestion are cattle manure and agricultural residues. The biogas plants can be broadly be classified into two types such as fixed dome plants and floating gas-holder digester plants. Biogas is generally used for cooking by supplying the gas through pipes to households from the plant. Biogas has been effectively used as a fuel in industrial high compression spark-ignition engines. Biogas can be effectively used as fuel in water heaters by completely removing H_2S during the supply of the gas.

1.3.1 FIXED DOME BIOGAS DIGESTER PLANT

Generally, the structure of a biogas plant is composed of brick and cement having five components such as mixture tank, inlet tank, digester, outlet tank, overflow tank (Sims, 2007). The mixing tank is located above ground level, whereas the inlet tank is kept underground into an inclined chamber, and the inlet tank is opened just below the large digester tank. The ceiling of the tank is kept as a dome-shaped structure on which a long pipe is connected using an outlet valve for supplying the biogas. The digester opens from the bottom side of an outlet chamber, and the outlet chamber opens from the top side into a small overflow tank. The typical schematic of a fixed dome type biodigester plant is shown in Figure 1.2.

Firstly, the biomass available in different forms is mixed thoroughly with an equal amount of water in a mixer tank to make a slurry to feed the digester through the inlet chamber. Induction of the slurry into the digester tank is stopped after partial filling and the biogas plant is intentionally left for two months without any use. Thereafter, the remaining anaerobic bacteria in the slurry decompose the biomass (fermentation) in water for several weeks to generate a high quantity of biogas, which is collected in the dome above the digester tank. The more the collection of biogas, the more is the exerted gas pressure to force the used slurry into the outlet chamber. This slurry overflows from the outlet chamber after filled and then removed manually for making it manure. A gas valve as shown in Figure 1.2, controls the supply of biogas through a pipeline connected to the dome. To maintain a continuous supply of biogas through the piping system, it is recommended to feed the slurry continuously to the digester.

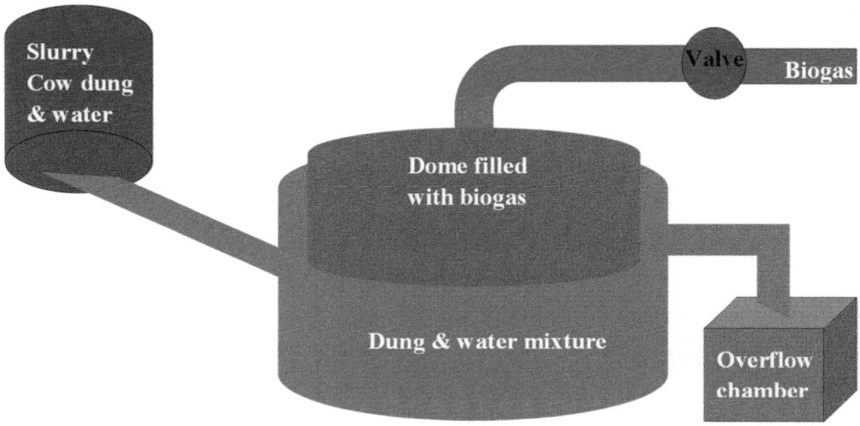

FIGURE 1.2 Fixed dome type biogas digester plant.

1.3.2 FLOATING DOME BIOGAS DIGESTER PLANT

A common movable or floating dome type biogas digester plant or gobar gas plant is shown in Figure 1.3. First, the raw materials like cattle manure and water are mixed properly in the input tank, then the mixture is allowed to enter inside the large digester tank, where the digestion takes place biologically or anaerobically (Desai and Palled, 2013). The biogas generated from the digester moves upwards and stored in the gas holder tank or dome (Castano et al., 2014). The movement of the gasholder is restricted up to a particular level. With the collection of more and more biogas, greater pressure is likely to be exerted in the slurry. Then the used slurry is forced from the top of the inlet chamber into the outlet chamber. When the spent slurry fills the outlet chamber, then excess quantity is forced into the overflow tank through the outlet pipe, which is later on used as the best manure for the plants. A bent pipe is fitted at the top of the dome for the flow of the biogas, which is controlled by opening and closing a gas valve as shown in Figure 1.3. Once the production of biogas begins, a continuous supply of gas can be ensured by regular removal of spent slurry and induction of fresh slurry into the inlet chamber. Practical use of such biogas for cooking alone since it requires the waste material like cattle manure and other household wastes (Sathish et al., 2017).

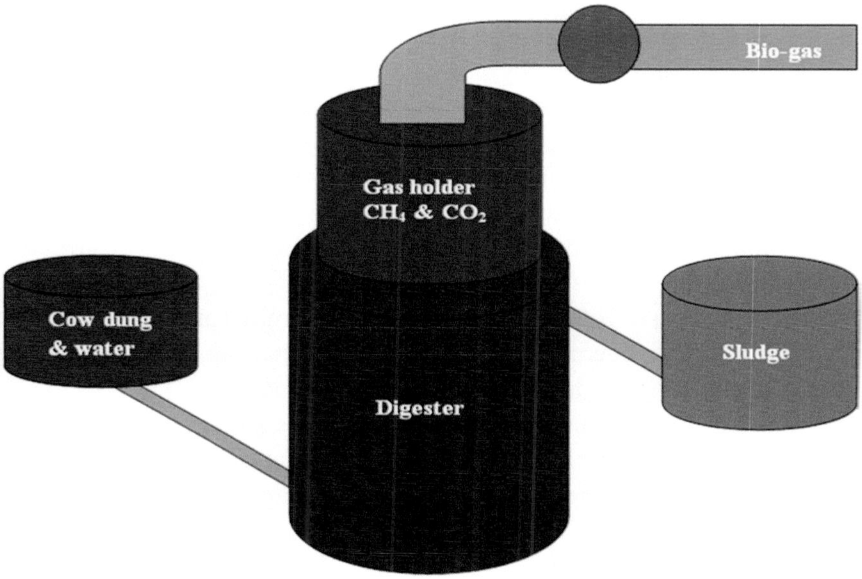

FIGURE 1.3 Floating dome type biogas digester plant.

1.4 ADVANTAGES AND CHALLENGES OF BIOGAS PLANTS

The following are some advantages for the biogas plants (Zheng et al., 2014):

(i) It reduces the consumption of fossil fuels.
(ii) It generates clean fuel for renewable energy and partially controls air pollutions.
(iii) It uses nutrient-rich manure to be used for the plants.
(iv) It controls water pollution by decomposition of animal manure and solid wastes.
(v) The biosludge obtained after anaerobic digestion can be used as an organic fertilizer in agriculture.

There are certain limitations for small biodigester plants as follows (Rasi et al., 2007):

(i) Small communities cannot generate enough solid waste, cattle manure, and nitrogenous organic materials to feed the installed biodigester plant. Therefore, the installation, operation, and maintenance of biodigesters could seem expensive for smaller communities.
(ii) The working temperature for a biodigester is 35–40°C, which is difficult to achieve in the colder weather.
(iii) There is also a chance of rusting of steel drum of the digester.
(iv) All the disease-causing pathogens are not fully destroyed by the anaerobic environment, acidic pH of the slurry and insufficient heating.
(v) The operators of the biodigestor must be familiar and knowledgeable for its safe operation, troubleshooting, repairing, and maintenance since the gas is highly flammable and obnoxious.
(vi) Some of the household feedstocks could contain bleach and disinfectants. Such compounds might pose antimicrobial action, thus inhibiting the methanogenesis in the biodigester.

1.5 CONCLUSIONS

Biogas is considered a fresh source of energy, generally used as a domestic biofuel for cooking, heating, street lighting as well as combined heat and power generation. It is a substitute for compressed natural gas for vehicles and the generation of electricity. Biogas also demonstrates a high calorific value with clean and smokeless emissions upon combustion. It does not retain any residue and is economical, combustible, and has convenient ignition temperature. Anaerobic digesters are mature biodigesters with many promising aspects in the rural sectors around the world. The research organizations should emphasize new and efficient methods to design and develop low-cost biogas plants. Hence, more effort is required to overcome the limitations of popularizing the biogas technology in rural areas for the benefit of the people. The governments play a significant role in forming legal frameworks by introducing various schemes in education, technology, and simultaneously enhancing the awareness for biogas production, thereby providing more subsidies.

KEYWORDS

- **biomass**
- **crude oil**
- **fossil fuel**
- **liquefied petroleum gas**
- **renewable energy**

REFERENCES

Andrea, S., & Fernando, R., (2012). Identifying, developing, and moving sustainable communities through renewable energy. *W. J. Sci. Tech. Sus. Dev., 9*, 273–281.

Baçaoui, A., Yaacoubi, A., Dahbi, C., Bennouna, J., & Mazet, A., (1998). Activated carbon production from Moroccan olive wastes-influence of some factors. *Environ. Technol., 19*, 1203–1212.

Bessou, C., Fabien, F., Benoit, G., & Bruno, M., (2009). Biofuels, greenhouse gases and climate change. *Sustain. Agric.*, 6–15.

Bhatia, L., Sarangi, P. K., & Nanda, S., (2020). Current advancements in microbial fuel cell technologies. In: Nanda, S., Vo, D. V. N., & Sarangi, P. K., (eds.), *Biorefinery of Alternative Resources: Targeting Green Fuels and Platform Chemicals* (pp. 477–494). Springer Nature, Singapore.

Castano, J. M., Martin, J. F., & Ciotola, R., (2014). Performance of a small-scale, variable temperature fixed dome digester in a temperate climate. *Energies, 7*, 5701–5716.

Desai, S. R., & Palled, V. K., (2013). Performance evaluation of fixed dome type biogas plant for solid-state digestion of cattle dung. *Karnataka J. Agri. Sci., 26*, 103–106.

Gong, M., Nanda, S., Romero, M. J., Zhu, W., & Kozinski, J. A., (2017). Subcritical and supercritical water gasification of humic acid as a model compound of humic substances in sewage sludge. *J. Supercrit. Fluids, 119*, 130–138.

Hall, O., & Scrase, J., (1998). Will biomass be the environmentally friendly fuel of the future? *Biomass Bioenergy, 15*, 357–367.

Herold, M., Carter, S., Avitabile, V., Espejo, A. B., Jonckheere, I., Lucas, R., Sanz, M. R. E., et al., (2019). The role and need for space based forest biomass related measurements in environmental management and policy. *Surveys Geophys., 40*, 757–778.

Nanda, S., Azargohar, R., Dalai, A. K., & Kozinski, J. A., (2015). An assessment on the sustainability of lignocellulosic biomass for biorefining. *Renew. Sust. Energy Rev., 50*, 925–941.

Nanda, S., Dalai, A. K., Gökalp, I., & Kozinski, J. A., (2016). Valorization of horse manure through catalytic supercritical water gasification. *Waste Manage., 52*, 147–158.

Nanda, S., Mohammad, J., Reddy, S. N., Kozinski, J. A., & Dalai, A. K., (2014). Pathways of lignocellulosic biomass conversion to renewable fuels. *Biomass Convers. Bioref., 4*, 157–191.

Nanda, S., Rana, R., Hunter, H. N., Fang, Z., Dalai, A. K., & Kozinski, J. A., (2019). Hydrothermal catalytic processing of waste cooking oil for hydrogen-rich syngas production. *Chem. Eng. Sci., 195*, 935–945.

Nanda, S., Rana, R., Vo, D. V. N., Sarangi, P. K., Nguyen, T. D., Dalai, A. K., & Kozinski, J. A., (2020). A spotlight on butanol and propanol as next-generation synthetic fuels. In: Nanda, S., Vo, D. V. N., & Sarangi, P. K., (eds.), *Biorefinery of Alternative Resources: Targeting Green Fuels and Platform Chemicals* (pp. 105–126). Springer Nature.

Nanda, S., Reddy, S. N., Vo, D. V. N., Sahoo, B. N., & Kozinski, J. A., (2018). Catalytic gasification of wheat straw in hot compressed (subcritical and supercritical) water for hydrogen production. *Energy Sci. Eng., 6*, 448–459.

Okolie, J. A., Nanda, S., Dalai, A. K., & Kozinski, J. A., (2021). Chemistry and specialty industrial applications of lignocellulosic biomass. *Waste Biomass Valor*, 12, 2145–2169.

Omer, A. M., & Yemen, F., (2003). Biogas energy technology in Sudan. *Renew. Energy, 28*, 499–507.

Omer, A. M., (2009). Energy use and environmental impacts: A general review. *J. Renew. Sustain. Energy, 1*, 53101.

Omer, A. M., (2012). Applications of biogas: State-of-the-art and future perspective, *Blue Biotech. J., 1*, 335–383.

Omer, A. M., (2012). Biomass energy resources utilization and waste management. *Agri. Sci., 3*, 16848.

Omer, A. M., (2015). Utilisation of biomass for renewable bioenergy development. *Am. Res. J. Biosci., 2*, 3–27.

Parakh, P. D., Nanda, S., & Kozinski, J. A., (2020). Ecofriendly transformation of waste biomass to biofuels. *Curr. Biochem. Eng., 6*, 120–134.

Rana, R., Nanda, S., Reddy, S. N., Dalai, A. K., Kozinski, J. A., & Gökalp, I., (2020). Catalytic gasification of light and heavy gas oils in supercritical water. *J. Energy Inst., 93*, 2025–2032.

Rasi, S., Veijanen, A., & Rintala, J., (2007). Trace compounds of biogas from different biogas production plants, *Energy, 32*, 1375–1380.

Rasslavicius, L., Grzybek, A., & Dubrovin, V., (2011). Bioenergy in Ukraine – possibilities of rural development and opportunities for local communities. *Energy Policy, 39*, 3370–3379.

Sarangi, P. K., & Nanda, S., (2018). Recent developments and challenges of acetone-butanol-ethanol fermentation. In: Sarangi, P. K., Nanda, S., & Mohanty, P., (eds.), *Recent Advancements in Biofuels and Bioenergy Utilization* (pp. 111–123). Springer Nature, Singapore.

Sarangi, P. K., & Nanda, S., (2019). Recent advances in consolidated bioprocessing for microbe-assisted biofuel production. In: Nanda, S., Sarangi, P. K., & Vo, D. V. N., (eds.), *Fuel Processing and Energy Utilization* (pp. 141–157). CRC Press, Florida.

Sarangi, P. K., & Nanda, S., (2020). Biohydrogen production through dark fermentation. *Chem. Eng. Technol., 43*, 601–612.

Sarangi, P. K., & Sahoo, H. P., (2010). Enhancing the rate of ferulic acid bioconversion using glucose as carbon source. *J. Am. Sci., 6*.

Sarangi, P. K., Nanda, S., & Vo, D. V. N., (2020). Technological advancements in the production and application of biomethanol. In: Nanda, S., Vo, D. V. N., & Sarangi, P.

K., (eds.), *Biorefinery of Alternative Resources: Targeting Green Fuels and Platform Chemicals* (pp. 127–140). Springer Nature.

Siang, T. J., Roslan, N. A., Setiabudi, H. D., Abidin, A. Z., Nguyen, T. D., Cheng, C. K., Jalil, A. A., et al., (2020). Recent advances in steam reforming of glycerol for syngas production. In: Nanda, S., Vo, D. V. N., & Sarangi, P. K., (eds.), *Biorefinery of Alternative Resources: Targeting Green Fuels and Platform Chemicals* (pp. 399–426). Springer Nature.

Sims, R. H., (2007). *Not Too Late: IPCC Identifies Renewable Energy as a Key Measure to Limit Climate Change.* Renewable energy world. https://www.renewableenergyworld.com/2007/07/01/not-too-late-ipcc-identifies-renewable-energy-as-a-key-measure-to-limit-climate-change-51487/#gref (accessed on 25 June 2021).

Singh, A., (2008). Biomass conversion to energy in India: A critique. *Renew. Sust. Energy Rev., 14*, 1367–1378.

Taherzadeh, M. J., & Karimi, K., (2008). Pretreatment of lignocellulosic wastes to improve ethanol and biogas production: A review. *Int. J. Mol. Sci., 9*, 1621–1651.

Zheng, Y., Zhao, J., Xu, F., & Li, Y., (2014). Pretreatment of lignocellulosic biomass for enhanced biogas production. *Prog. Energy Combust. Sci., 42*, 35–53.

CHAPTER 2

Characteristics, Parameters, and Process Design of Anaerobic Digesters

APOORVA UPADHYAY,[1] NIDHI PAREEK,[2] and VIVEKANAND VIVEKANAND[1]

[1]*Center for Energy and Environment, Malaviya National Institute of Technology, Jaipur, Rajasthan, India*
E-mail: vivekanand.cee@mnit.ac.in (Vivekanand Vivekanand)

[2]*Department of Microbiology, Central University of Rajasthan, Ajmer, Rajasthan, India*

ABSTRACT

The anaerobic digester is widely used in wastewater treatment plants owing to its good efficacy, low energy usage, and renewable energy production. In this chapter, the merits and demerits of anaerobic reactions were shown, along with the analysis of key features of an anaerobic digester system. For high productivity in wastewater treatment, anaerobic digestion system parameters such as organic loading rate, pH, hydraulic retention time, temperature, and sludge retention period are adopted to understand the best environments for bacteria living, growing, and multiplying. Also described are the internal machinery boosting sludge retention time to increase the solid movement. This is categorized into longitudinal and transverse inward components.

2.1 INTRODUCTION

Anaerobic degradation, a likable option for waste management practices which can accomplish energy saving as well as pollution reduction. In the absence of oxygen, anaerobic mortification or fermentation includes the breakage of biological substances through an intensive accomplishment of varied microorganisms. Biological processes are mainly used in wastewater treatment and several biological recovery techniques are now available, and positive results have been obtained in the recovery of complex organic matter (Mittal, 2006, Wijetunga et al., 2010). Anaerobic approaches are primarily called an appropriate treatment alternative due to reduced energy demands and small amounts of sludge production as combined with anaerobic and aerobic approaches. Anaerobic approaches were often particularly difficult when treating different industrial runoff, including toxic chemicals or also low amounts of household runoff (Seghezzo et al., 1998; Aiyuk et al., 2006). The promise of achieving ecological conservation and carbon recovery, anaerobic process systems and anaerobic tanks has achieved substantial consideration (Seghezzo et al., 1998; McCarty et al., 1964; Chong et al., 2012). The anaerobic digestion (AD) process is an appropriate and important source of energy. A typical digester comprises a digestion compartment, a lid, an inlet, a biogas outlet and a slurry outlet. The biogas contained by the dome moves under heat through the vent, where it gradually becomes an origin of energy.

The documented usage of the anaerobic reactor probably began in 1859 when a leper colony built the first anaerobic digester in Bombay, India (Meynell, 1976). In Exeter, England, during 1895, the idea was created for the sewer gas destructor lamp in where a septic tank was being cast-off to manufacture H_2, a type of gaslighting. Anaerobic digestion received significant scientific recognition in the 1930s (Reddy, 2011). The first type of anaerobic bioreactors (AB) was reflected in the septic chamber. The up-flow anaerobic sludge blanket (UASB) system, developed in the late 1970 s was second-generation AB (Lettinga et al., 1980; Kato et al., 1994). Initially, the UASB reactors acted like a vacant chamber where a three-phase separator was inserted to keep the sludge from washing out of granular materials.

The third generation of AB was generated focused on the internal circulation (IC) and extended granular sludge bed (EGSB) reactor. For example, for wastewater that passed through the sludge bed, the advanced frequency of up-flow speed is expected. Amplified fluctuation enables fractional enlargement (fluidization) of the coarse sludge layer, refining the interaction between drainage and sludge, as well as increasing the separation from the sludge beds of tiny inactive suspended particles. The increased flow rate is accomplished either by the use of broad reactors or by the adoption of effluent recycling (Seghezzo et al., 1998; Chen et al., 2010; Chen et al., 2011; Chen et al., 2012). Operational factors such as the speed of upward flow, solids retention time (SRT) and hydraulic retention time (HRT) are the two most significant variables for the cultivation of anaerobic digester by microorganisms. (Seghezzo et al., 1998; Chen et al., 2010; Chen et al., 2011; Chen et al., 2012). This work also defines the anaerobic cycle's advantages and disadvantages. This chapter discusses important operational factors, internal machinery and its impact upon anaerobic bioreactors' efficiency in wastewater treatment.

2.2 APPLICABILITY OF ANAEROBIC PROCESS

Anaerobic systems for the treatment of sewage, particularly UASB reactors, have evolved in complexity because of increased awareness, holding a desirable place in many countries with tropical temperature conditions. The reception moved after a period of denial to the present era of universal recognition, which continued until the early 1980 s. This recognition has also contributed to project construction and the introduction of sewage degradation plants, which have significant technical issues. The following discussion seeks to include knowledge related to the concepts, architecture, and application of anaerobic systems for treatment of sewage, with a focus on anaerobic UASB and anaerobic filters. Both biological substances can be processed in theory by a more productive and cost-effective method, namely the anaerobic method because the waste is readily biodegradable.

Anaerobic digesters have been used largely in the disposal of solid waste, including farm waste, animal manure, water treatment system sludge, and industrial waste, and millions of anaerobic digesters are known to have been installed around the world for this reason. Anaerobic digestion has also been commonly used in both developed and emerging countries in the management of effluents from the livestock, food, and beverage industries. A significant improvement in the usage of anaerobic technologies has also been confirmed about the treatment of domestic sewage in warm-climate areas, especially by UASB-type reactors. The use of anaerobic technology in this situation relies even more on the water temperature owing to the weak development of anaerobic microorganisms at temperatures below 20°C and the failure to heat the reactors. This is since domestic waste is additionally reactive than commercial effluent, subsequent in higher emission amounts of methane gas, making its usage as a heat energy source uneconomical. Thus, anaerobic domestic sewage treatment is much more appealing to the nations of the tropical and subtropical world, which are mainly developing nations. Figure 2.1 gives a better representation of a few merits of the anaerobic digestion process compared to aerobic treatment, especially in terms of methane gas production and low solid growth.

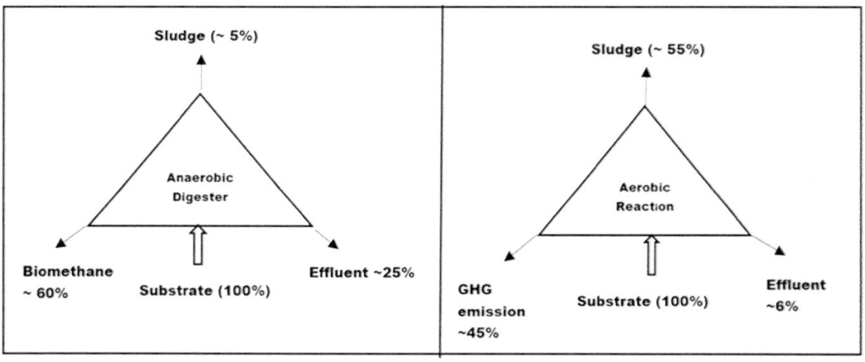

FIGURE 2.1 A comparison of anaerobic and aerobic processes.

Anaerobic digestion is a corroborated and effective processing technique for biomethane that can be used to generate green heat and electricity and a compost such as production. The usage of anaerobic microorganisms to convert biological substance or chemical oxygen demand (COD) in biomethane, is the theory of anaerobic treatment. The beginning of anaerobic granules on the hard surfaces obey the same laws as regards the production of bacterial biofilm. Inert carriers play a significant positive function in granulating. The methanogenic bacteria, until the channels are in equilibrium, use acidic intermediator products as soon as they come through (Hulshoff et al., 2004). Nevertheless, in the absence of sufficient concentrations of methanogenic bacteria or if unfavorable environmental factors slow down, the acids are not consumed as quickly as acid formers produce them, thereby increasing the concentration of volatile acids. An increase in acid demand thus means that the formers of methane will not follow the acid formers (McCarty et al., 1964). Table 2.1 indicates both merits and demerits of anaerobic digestion strategies.

TABLE 2.1 Merits and Demerits of the Anaerobic Digestion Processes

Merits	Demerits
• Less solid development, i.e., around 3–5 times less than in aerobic processes	• A significant range of substances are subject to inhibition by anaerobic microorganisms
• High energy use, commonly correlated with a powerful pumping station, which results in very high running costs	• In the absence of suitable seed sludge, plant start-up may be sluggish
• Fewer land demands	• It normally requires some sort of post-treatment
• Less construction costs	• Anaerobic digestion biochemistry and microbiology are complicated, and further studies are still needed
• Methane has a high heating value	• Possible generation of unpleasant smells even though they are controllable
• Possibility of storing the biomass for many months, without reactor feeding	• Possible production of bad looking effluents
• Tolerance against heavy organic feed	• Insufficient degradation of ammonia, phosphorus
• Many applications	• A significant range of substances are subject to inhibition by anaerobic microorganisms
• Less intake of nutrients	• In the absence of suitable seed sludge, plant startup may be sluggish

Source: Lettinga et al. (1996); Chernicharo and Campos (1995); von Sperling (1995).

2.3 TYPE OF BIOREACTOR FOR ANAEROBIC DIGESTION

The nature of systems for handling biological wastewater is in the capacity of the microorganisms concerned to use biodegradable organic compounds to turn them into by-products that can be extracted from the treatment network. The produced by-products may be in the form of solid (biosludge), liquid (water) or gaseous (CO_2, CH_4, etc.); The potential for using the organic compounds in any method employed, aerobic or anaerobic, which is dependable on the microbial behavior of the biological compounds contained in the environment. Until recent times it was viewed as economic and problematic to use anaerobic approaches for liquid effluent treatment.

The diminished rate of growth of the anaerobic biomass, specifically of the methanogenic bacteria, is delicate process control since the system recovery is slow if the atmosphere is adverse. High-rate devices have been built with the advancement of anaerobic-treatment science. These are distinguished by their capacity to maintain significant volumes of high-activity biomass, particularly though short hydraulic retention periods are added. Moreover, it maintains a strong holding time for solids, with large hydraulic loads sometimes applied to the network. The result is limited reactors with volumes below traditional anaerobic tanks but retention of the high amount of sludge stability. For simplicity, those are grouped into two broad categories, as seen below:

(i) Traditional systems (e.g., anaerobic ponds, sludge digesters and septic tanks).
(ii) High energy networks with contact growth (e.g., fixed bed reactor, rotating bed reactor and fluidized bed reactor).
(iii) High energy networks with discrete growth (e.g., two-stage reactors, up-flow anaerobic sludge blanket, expanded gritty bed reactors and reactors with inner recirculation).

2.4 PROPERTIES OF ANAEROBIC DIGESTER

2.4.1 NON-HOMOGENEOUS SYSTEM

The anaerobic reactor works in a nonhomogeneous system, which means three stages of anaerobic treatment are performed, namely firm (sludge), fluid, and gas (methane). The solid stage consists of sludge pellets of around 0.5–2 mm in diameter (Hulshoff et al., 2004), which exist in the lower part of the reactor. Sludge granules play an important role in the efficient activity of UASB and EGSB. Sludge granules are the group of microorganisms produced in a wastewater system without any supporting matrix and in an atmosphere of continuous up-flow in the course of the flow of wastewater.

Water is pumped into a bottom section of the reactor by a sludge layer, although the upper part includes a separation system of three levels (i.e., solid, water, and gaseous). The most characteristic feature of the UASB reactor is the three-phase separation system. It enables biogas processing and provides internal sludge recovery by separating adherent biogas from growing particulate matter (Chen et al., 2012).

2.4.2 TIME INSTABILITY

Anaerobic digestion is used in wastewater treatment and consists of intricate biological matter. With the increase in output cycles and amounts, it gets more difficult. Different industrial processes that operate under various circumstances (e.g., time, temperature, and pH) with a broad variety of several seasons in a year had produced a task on AD. Because of the intrinsic shortcomings of anaerobic process technologies, it is important to pay attention to improving its demerits, and thus challenge designers and engineers (Chong et al., 2012). As a consequence of supporting anaerobic digestion disposal in a consolidated sewage treatment plant (STP), several regional sewage treatment plants work separately in local cities; sewage sludge with more amount of solids is advantageous as it decreases the time and costs needed to transport the sludge from additional sewage treatment plant.

2.4.3 SPACE INSTABILITY: ANAEROBIC REACTION PROCESS

Microbial consortia accomplish the translation of composite biological matter into biomethane, and each performs a particular role in the digestion cycle. Together, they complete a series of stages in the processing of organic matter into biogas. Figure 2.2 demonstrates the main anaerobic digestion cycle stages (Abbasi et al., 2012). Hydrolysis is the first step. It is particularly essential for the anaerobic digestion process, as the fermentative microorganisms cannot use polymers directly. The insoluble, versatile biomaterials like cellulose also become dissoluble compounds like proteins, amino acids and fatty acids. The composite polymers are hydrolyzed to monomer by enzymes produced by hydrolytic microorganisms, viz. amylases, proteases, lipases, cellulases, etc., thereby becoming accessible to other microorganisms (Madigan et al., 2008; Kayhanian et al., 1995). In the hydrolysis reaction, a complex polysaccharide is fragmented into a simple sugar, e.g., glucose (Madigan et al., 2008, Kayhanian et al., 1995).

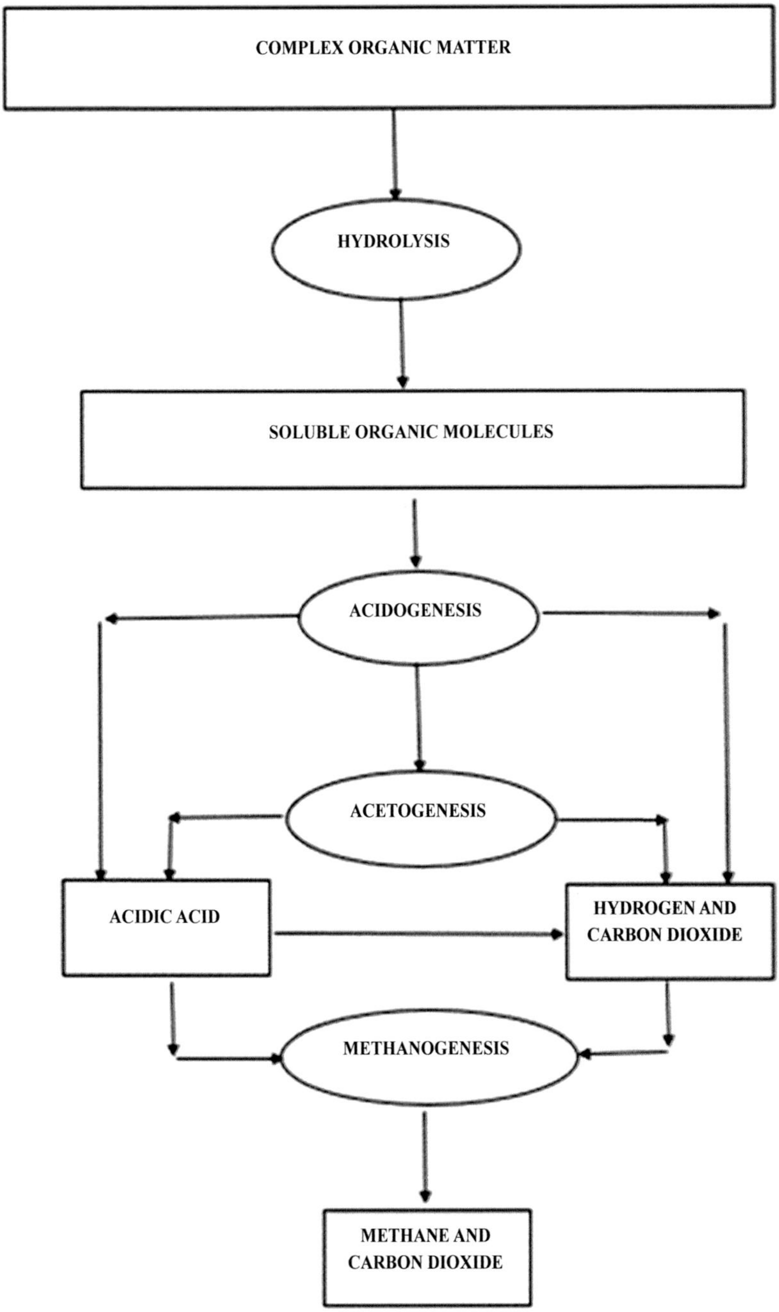

FIGURE 2.2 The main steps in the cycle of anaerobic digestion.

In the second stage of acid genesis (fermentation), acid bacteria turn the products viz. sugars and amino acids, are converted into CO_2, H_2, NH_3, and organic acids. In acidogenesis reaction, ethanol, acetic acid, butyric acid, and propionic acid are the major products. In a controlled environment, much of the organic matter is converted into readily available substrates for methanogenic microorganisms (acetate, H_2 and CO_2), but a small portion (about 30%) is transformed into short-chain fatty acids or alcohols (Kangle et al., 2012). H_2, CO_2 and acetic acid from these materials will bypass the acetogenesis stage and directly convert into CH_4 by the methanogenic bacteria in the final stage. The three common reactions of acid regeneration where fructose, acetic acid and propionate are transformed from glucose (Ostrem et al., 2004; Kangle et al., 2012).

In the third step, CH_3COOH, H_2, and CO_2 are formed from the effects of the chemical compounds and organic acids, which is obtained from the second process. The reduction of propionate in acetate is seen at small concentrations of H_2. Throughout the third stage the anaerobic digestion cycle, glucose, and ethanol are both transformed into acetate (Ostrem, 2004). The compounds produced in acetogenesis are attributable to a variety of various microorganisms, viz. *Syntrophobacter wolinii* and *Sytrophomonos wolfei*, which are propionate decomposer and butyrate decomposer, respectively. Numerous acid formers are *Actinomyces, Lactobacillus, Peptococcus anaerobius* and *Clostridium* spp. (Kangle et al., 2012).

Methanogenesis is the final stage where CH_4 is generated by methanogens categorized as bacteria and archaea, either through the cleavage of acetic acid molecules to generate CO_2 and CH_4, or through the reduction of H_2 and CO_2. CH_4 production is higher from CO_2 reduction, but a small concentration of H_2 in digesters contributes to an acetate reaction becoming the main source of CH_4 (Kayhanian et al., 1995). The methanogenic bacteria are *Methanobacterium, Methanobaccillus, Methanocococcus, and Methanosarcina.* Methanogens can also be classified into two groups, i.e., the users of acetate and H_2/CO_2 (Kayhanian et al., 1995; Berger et al., 2012):

The problems facing anaerobic digestion such as low yield of CH_4 and process instability, preventing widespread implementation of this technology. The primary cause of anaerobic digester upset or failure is a wide variety of inhibitory substances, as they exist in a huge amount in waste. A substantial investigation was made to define the inhibition mechanism and control factors. The methane-producing bacteria are purely anaerobic and they are toxic to even small amounts of oxygen (McCarty et al., 1964). Material common to anaerobic species is nitrogen (Mah et al., 1978). The ammonia concentration must be preserved above 45–65 mg N/L to avert a drop in biomass movement (Takashima et al., 1989). Moreover, high concentrations of ammonia can result in inhibition of anaerobic digestion (Hansen et al., 1998; Kayhanian et al., 1999; Sung et al., 2003). Nevertheless, better reactor design should reduce spatial uncertainty as soon as possible. For illustration, by adding three inner sections that help to maintain space stability, the anaerobic compartmentalized reactor (CAR) (Zheng et al., 2012) has been separated into a delivery region, a reaction zone and a splitting field. Consequently, the CAR demonstrates the enormous capacity for its use.

2.5 SYSTEM PARAMETERS IN ANAEROBIC DIGESTION

Degradation in the anaerobic handling of undesirable components or contaminants relies on many factors. Reactor operating conditions and essential features are the key parameters that include the distribution of particle size. Such parameters and outcomes are discussed here.

2.5.1 TEMPERATURE

Temperature is a significant physical phenomenon that influences both water acceptability and water chemistry and water treatment (Adrian et al., 2008). Anaerobic bacteria are categorized as temperature groups depending on equilibrium temperature; mesophiles live at mesophilic temperatures between 30°C and 40°C, while thermophiles are known to be the first microorganisms to thrive at high temperatures between 50°C and 65°C. (Agler et al., 2010). Heat also influences to some extent all wastewater treatment systems, such as (Adrian et al., 2008; Gao et al., 2011).

- Biological waste treatment: Coldwater diminishes high-filter performance by around 30%. Low temperatures impede nitrification rather than exclusion of biological oxygen demand (BOD).
- Breakdown: The average solid survival period fluctuates from 2 days (at 35°C) to 10 days (at 20°C). The digester's high-temperature demand is determined by outside temperature.

- Microbial development: The temperature affects the richness and range of microbial communities.

Temperature influences particle removal by affecting the viscosity of wastewater and the transfer of organic matter (Mahumoud et al., 2003). Because water viscosity is biologically associated with temperature, temperature changes can affect microorganisms' behavior through both physical and physiological means (Winkler et al., 2012). Due to the biological association of water viscosity with temperature, temperature fluctuations may affect the behavior of minuscule creatures through physiological or physical means (Mahumoud et al., 2003). The rapid temperature change can result in a variation in the chemical and physical characteristics of wastewater, which may have a significant impact on the anaerobic digestion (Lettinga et al., 2001).

The temperature usually affects the intracellular and extracellular functionalities of the bacteria and serves as a transition cycle driver. Initialization may take longer at low temperatures but can be easily carried out with digested sludge inoculation of the reactor. Anaerobic treatment of untreated domestic wastewater (COD of around 550–650/g^3) can be carried out at around 12–18°C in UASB reactors (Seghezzo et al., 1998). Furthermore, the probability of anaerobic degradation of diluted dairy wastewater at around 12°C in OLRs up to around 2.5 kg COD/m^3/d can improve the efficiency of oxygen demand removal by more than ~85% (Katarzyna et al., 2013).

At higher temperatures, the reaction speeds continue much quicker, thereby creating more effective processing and smaller tank sizes. Treatment often occurs even quicker at thermophilic levels (around 50–65°C), so high-temperature digestion conditions provide several benefits, such as advanced metabolic speeds, developed individual growth rate, yet also advanced mortality rates relative to mesophilic bacteria, as well as extra high temperature needed to sustain these levels. Then, most care methods are built to operate inside or below the mesophilic spectrum (McCarty et al., 1964; Mashad et al., 2004; Ge et al., 2011).

2.5.2 pH VALUE

The pH is an indicator of the strength of a liquid's neutral or acidic state, a measure of a solution's acidity. Throughout wastewater treatment facilities, methanogens are typically most involved inside the neutral pH spectrum (7.0) (Florencio et al., 2011). In most species, the normal pH range is 6.0–9.0 (Adrian et al., 2008); digestion may continue within the range but with reduced performance. Upon changing pH to equilibrium, the cell mass, which was inhibited at pH ~8.9, resumed activity (Fang et al., 1998). The methanogenic bacteria may become very toxic in generated acidic environments. For this cause, it is necessary not to allow the pH to drip under 6.3 for a substantial period. Since this function is so critical, the device will then regulate the pH. The pH stays between 7.2 and 8.2 until CH_4 output stabilizes (Abbasi et al., 2012). McCarty et al. (1964) indicated the optimum pH range in anaerobic treatment is around 7.2–7.5, although it can continue well at a pH range of 6.6–7.6.

2.5.3 *HYDRAULIC RETENTION TIME*

HRT, also is known as hydraulic residence period or hydraulic retention time is a calculation of the total duration of time a soluble material stays in a biodigester being built. Hydraulic holding time is the amount separated by the powerful flow rate of the aeration tank:

$$\mathrm{HRT}\left[\mathrm{d}\right]=\frac{\text{Volume of Aeration tank}[\mathrm{m}^3]}{\text{Influent Flow rate}[\frac{\mathrm{m}^3}{\mathrm{d}}]}=\frac{V}{Q} \tag{2.1}$$

Where HRT is the hydraulic retention time (d) measured in hours or days, V is the volume of the aeration tank/reactor volume (m^3), and Q is the flow rate (m^3/d).

2.5.4 *ORGANIC LOADING RATE*

Some reactors such as up-flow staged sludge bed (USSB), UASB, EGSB, and IC can accommodate reactors with high load with up to ~45 kg of COD/(m^3d) with the anaerobic upright sheet (Chen et al., 2010). In our laboratory, the spiral automatic circulation (SPAC) reactor organic charge rate (OLR) and volumetric biogas production (VBP) could exceed COD/(m^3d) up to 306 kg (Chen et al., 2010; Wang et al., 2009; Chen et al., 2008). Some reports have reported that the treatment ability of complex wastewater, such as potato corn, slaughterhouse, is to some degree that with an improvement in OLR in high-rate anaerobic reactors. Further rises for OLR will contribute to glitches due to operations such as flotation and excessive gas-liquid-solid (GLS) spray on the gas-fluid interface and to the aggregation of materials. Therefore, the quality of care worsens (Ruiz et al., 1997; Kalyuzhnyi et al., 1998). There was also a significant accumulation of biogas in the sludge layer, creating secure gas pockets that lead to the incidental lifting of layer parts and pulse-like gas eruption from this zone (Kalyuzhnyi et al., 1998; Elmitwalli et al., 1999). By changing the concentration of the variables and increasing the flow rate, the OLR can vary. Therefore, implies a change of the HRT and the flow rate, under these conditions OLR can be expressed as:

$$OLR = \frac{(Q^*COD)}{V} \tag{2.2}$$

where, OLR is the organic loading rate (kg COD/m^{3}d), Q is the flow rate (m^3/d), COD is the chemical oxygen demand (kg COD/m^3) and V is the reactor volume (m^3). The OLR can be minimized with the following reaction:

$$OLR = \frac{COD}{HRT} \tag{2.3}$$

When the efficacy of removal of solids in up-flow tanks is linked to the OLR, differentiating between these parameters is crucial. Because of this, OLR is an insufficient architecture parameter to ensure anaerobic reactors work well.

2.5.5 SOLID RETENTION TIME

Solid retention time (SRT) can influence the biochemical and physical properties of sludge (Halalsheh et al., 2005). The performance of UASB reactors depends primarily on the retention period of the sludge (SRT) (Varel et al., 1980), which is the main determinant of the ultimate hydrolysis and methanogenesis in the UASB system under certain temperature conditions (Jewell et al., 1987). At the prevailing conditions, the SRT should be long enough to have ample methanogenic operation. The loading rate, the fraction of the dominant suspended solid (SS), the SS deposition in the sludge layer, and the SS characteristics (e.g., biodegradability, composition, etc.),

(Seghezzo et al., 1998) decide the SRT. In SRT, methanogenesis begins for 7–12 days (at 25°C) and for 25–55 days (at 15°C) with average methanogenesis of 52% at 25°C and 25% at 20°C. The highest hydrolysis happens at ~72 days and which is up to 51% at 26°C and 25% at 20°C (Halalsheh et al., 2005). SRT and temperature have a significant effect on protein, starch, and lipid hydrolysis. The most important portion of protein, carbohydrate, and lipid degradation happens at cycle temperatures between 25°C and 35°C, respectively within the initial 15–10 days. (Mahmoud et al., 2004).

2.5.6 UP-FLOW VELOCITY

One of the main variables affecting the performance of up-flow reactors is the up-flow rate. A rise in the velocity of up-flow from 1.6 to 3.2 m/h resulted in a minor reduction in the efficiency of SS removal from 55% to nearly 50%, suggesting the role of adsorption and trapping (Zeeman et al., 1997).

2.5.7 PARTICLE SCALE DISTRIBUTION

The distribution of particle scale of a fluid distributed as liquid, grainy substance or particles is a set of standards or a mathematical equation that determines the relative number of particles present according to scale, usually by mass. The effluent content from classical filters is directly linked to the filtering media's unique thickness. Most findings suggest that reduced media scale allows removal more effective (Landa et al., 1997).

2.5.8 DESIGN OF ANAEROBIC REACTORS

2.5.8.1 ANAEROBIC FILTERS

The earliest tests of anaerobic filters date back to the end of the 1960 s. These have since been extended to a broad range of commercial wastewater sources and an advanced technique. A touch-device is simply the anaerobe up-flow screen, where water passes through a mass of biological solids trapped within the reactor. The retained biomass in the tank will come in three different forms:

- Partial biological layer coating fixed to medium outsides of packaging;
- Discrete biosolids treated at cracks in packaging sheet;
- Flocks or pellets in the lower container, beneath the litter pad.

Soluble organic compounds in water encounter the biomass and disperse across biofilm surfaces or the granule sludge. These are then transformed into transitional and finished goods, in particular CH_4 and CO_2. Automatic up-flow or downflow filter setups. These are then converted into intermediate and finished goods, in particular CH_4 and CO_2. Automatic up-flow or downflow filter setups. The packaging pad is completely immersed in up-flow filters. The downflow filters either may work in a submerged area or not submerged. They are normally closed, but when there is no doubt regarding the potential release of unwanted odors, they can be opened to. Anaerobic filters are mainly used for processing wastewater, thereby leading to overall operational health and efficiency of the treatment process.

Typically, the anaerobic filter effluent is well established, and organic matter has a comparatively low substance, whereas it is high in mineral salts. Its main use is for land usage, for drainage and for crop irrigation, as long as the issue of infection-causing microorganisms, mainly found in considerable quantities in the effluent, which is formed from domestic sewage treatment, is not overlooked. Throughout such situations, disinfection can become essential, and the normal existing protocols need to be implemented. The main difficulties of anaerobic filters are related to the risk of clogging of the interstices and the comparatively broad amount attributed to the region filled by the inert bag material. Anaerobic filters may have several types, heights, and distances, as long as the fluid is spread evenly around the surface. Broad shaped anaerobic filters are usually either rectangular or cylindrical. The tank diameters (or widths) range from about 7 m to 27 m, and the height ranges from about 4 m to 12 m. The capacities for the reactor vary from about 120 m^3 to 11,000 m^3. It is anticipated that the packaging media would take between 50% and 70% of the tank height from the overall volume of the reactor.

The industry offers numerous types of plastic packaging items, ranging from corrugated rings to corrugated plate covers. Usually, such plastic items have a typical surface area of between 100 m^2/m^3 and 200 m^2/m^3. While certain types of packaging media are more successful in saving biodiversity than others, the ultimate determination will be focused on the different limited circumstances, financial deliberations and organizational factors (Nanda and Berruti, 2021). The key optimum requirements for anaerobic filter packaging media are provided in Table 2.2. Many forms of materials have been used in biological reactors as packaging products, including marble, ceramic plates, oysters, and musk shells, calcareous, hollow tubes, rigid polyvinyl chloride (PVC) plates, granite, polyethylene balls, etc.

TABLE 2.2 Technology Specifications for Anaerobic Filters

Requisite	Objective
Be mechanically robust	Enable self-weight, supplemented to the biomass attaching to the substrate
Be eco-friendly and chemically inert	There is no connection between the layer and the microorganisms
Be light weighted	Evade the necessity for costly, hefty systems and enable comparatively higher filters to be mounted, which means a reduced area needed for network mounting
Have a wide devoted field	Allowing the incorporation of larger volumes of biological solids
Increased porosity	Allowing more free space for aggregating bacteria and growing the chance of obstruction
Enable speedy colonization of microorganisms	Reduce the reactor start-up period
A rugged and nonflat surface shape	Ensure a strong grip and huge absorbency
Have a less price	Making operation technologically and commercially feasible

2.5.8.2 UP-FLOW ANAEROBIC SLUDGE BLANKET BIOREACTORS

In tropical countries, particularly in Brazil, Colombia, and India, the usage of UASB reactors (Figure 2.3) to treat domestic sewage is already a fact. The positive experience in those countries is a good indicator of this form of reactor's capacity for inland sewage treatment. The anaerobic method through UASB reactors has many advantages over traditional aerobic processes, particularly when used in the climate with high-temperature environments. The device should have the following primary features in these situations:

- Lightweight, low-land necessity network;
- Low-slung energy usage;
- Adequate 65–75% COD and BOD separation efficiencies;
- Strong volume and strong waste properties of dewatering sludge.

Although there are many benefits of UASB reactors, there are also certain drawbacks or limitations:

- Possibility of the release of bad odors;
- Poor device ability to withstand toxic loads;
- Long machine start-up period;
- A post-treatment process is needed.

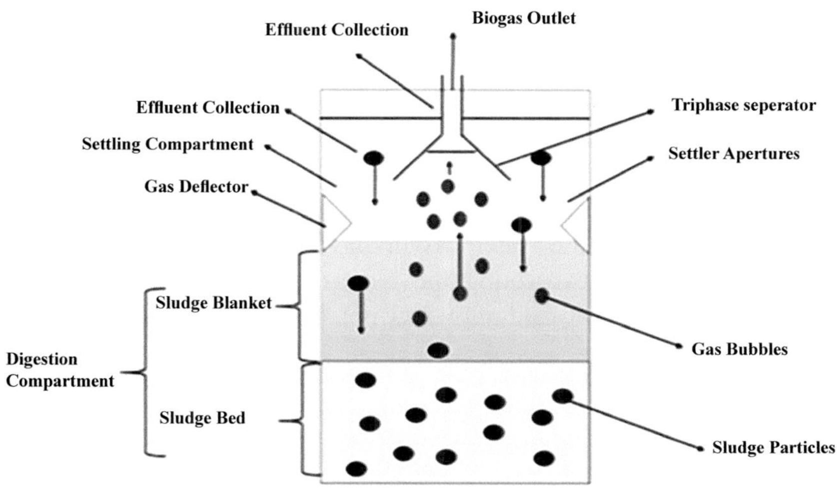

FIGURE 2.3 Schematic representation of UASB reactor.

The production of sulfur compounds and radioactive materials typically exists at very small levels in cases where the wastewater is primarily domestic, being treated well by the treatment method. Instead of the lack of radioactive elements and/or inhibitors, the device does not pose bad odor and breakdown issues when properly built, installed, and run. Node initialization may be sluggish (4 to 6 months), but only in such cases where no seed sludge is used. With well-founded start-up methodologies and the introduction of effective operating processes in the last three years, major changes have been made in the system's start-up phase, thereby the operational problems. In any event, it relies on the quality of biomass to be generated by the system and hence on the efficiency and efficacy of the treatment process. Nonetheless, the consistency of the traditionally processed effluent does not meet with the demands of most environmental discharge authorities, aside from the major advantages of the UASB reactors.

In recent years, there have been inadequate observations, which support an overview of collective phases of anaerobic diagnosis and post-treatment. Nonetheless, as Chernicharo et al. (2001b) reported, significant progress has been made lately. UASB reactors are very simple to construct and do not require any complex biomass processing devices or packaging media added. Given the cumulative expertise on UASB reactors, the architecture of such reactors still lacks simple, systematized instructions accessible to designers. The different design requirements and specifications for UASB reactors must be explicitly and sequentially articulated, enabling the chambers to be designed for reaction, sedimentation, and gas recovery. The basic principles that control the operation of UASB reactors are upward flow will ensure optimal contact with the biomass and the substratum's short circuits, providing sufficient processing cycles for the oxidation of the biological matter, the system would have a well-designed framework capable of properly extricating the biogas, the solids and the liquid, eliminating the first two and allowing protection of the remainder of the sludge.

2.5.8.3 SUPER HIGH-RATE ANAEROBIC BIOREACTOR

The biogas formed by anaerobic digestion cannot easily discharge granular sludge under high load conditions, decreases particle density and increases the stability of the soil. In biologically fluidized beds of laboratory size, slugging triggers paralysis of the reactor's mechanical activity. In reaction to this scenario, a new anaerobic spiral automatic circulation reactor (SPAC) is established (Wang et al., 2009) (Figure 2.4), OLR SPAC will achieve up to ~350 kg COD/m^3/d after more than two years, far higher than the existing high-rate anaerobic reactor output standard (Chen et al., 2008). Its key benefits are essentially avoiding the slugging tendency and maintaining reactor smooth activity using the spiral plate reactor's reaction power.

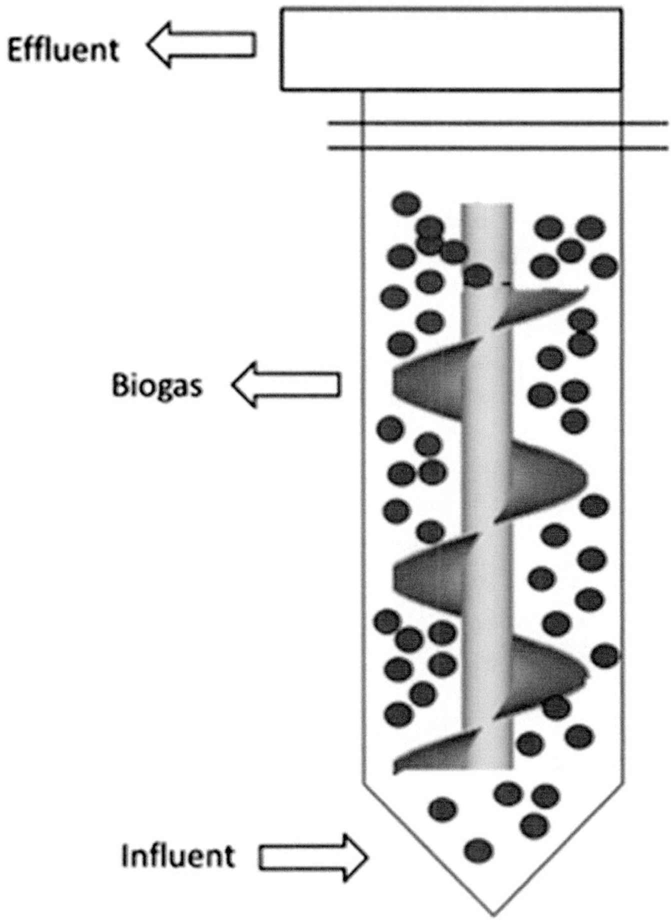

FIGURE 2.4 Schematic of spiral baffle.

2.5.9 INTERNAL COMPONENTS OF ANAEROBIC BIOREACTOR

The internal machinery performs a substantial part in improving the anaerobic digestion OLR, helping to boost the fluidization consistency, increasing the performance of the process by unraveling the gas effervesce from the sludge granules. The internal components that be divided into transverse internal parts, longitudinal internal and biofilm-packing materials in a biologically fluidized bed reactor.

2.5.9.1 TRANSVERSE INTERNAL MACHINERY

An utmost critical feature of transverse internal machinery is that the slop/sludge is successfully retained in an anaerobic bioreactor and which greatly increased fluidization, fracturing, and mass transfer. Several researchers in the early 1990 s placed components internally in the reactor by implementing the concept of three-stage parting in order to hold sludge effectively. Chelliapan et al. (2006) set a 45-phase separator baffle angle in the multistage reactor and efficiently kept the granular sludge (5850 gVSS/m^3). Several investigators set different internal sections in anaerobic bioreactor such that short flow is minimal and flow pattern is improved and mass transmission minimized.

TABLE 2.3 Comparison of the Fluidization Efficiency With and Without Hanging Inner Portion for Specific Conical Bottom Digesters

Small Setup	The small flux-region (%) of the total volume	
	Without inner components	**With inner components**
Flat bottomed head	32.6	20.4
Conical head ($\alpha = 25°$)	30.9	14.8
Conical head ($\alpha = 45°$)	28.6	10.7

Karim et al. (2007) examined the impact of bottom setup and a droopy baffle on the combining within a non-Newtonian sludge-filled gas lift digester. The data in Table 2.3 reveals that in the badly mixed region, the improvement in the digester bottom configuration resulted in a reduction of around 2–4%. The addition of a hanging baffle. However, lowered the percentage of badly mixed zones by 12%, 16% and 18%, respectively, for smooth, 25 and 45 low-level digesters. Thus, it is obvious that a hanging baffle combined with hopper bottom will greatly increase the mixing effectiveness of digesters within the gas lift.

2.5.9.2 LONGITUDINAL INTERNAL MACHINERY

Some work indicates that the potential production pattern of quantitative internal machinery is the biological incorporation of multigroup apparatuses to maximize the flow area and that the intermediate inhibition of the drug. The up-flow scale reactor (USSB) is in the sludge bed, Van Lier et al. (1996)

developed a sequence of separators for 3 phases. (Figure 2.5). Its output compared to that of an anaerobic up-flow sludge bed (UASB) reactor operating under the same operating conditions and the reaction zone in USSB is divided into several compartments, and each compartment may collect generated gas (Lens et al., 1998, Yanling, 1998).

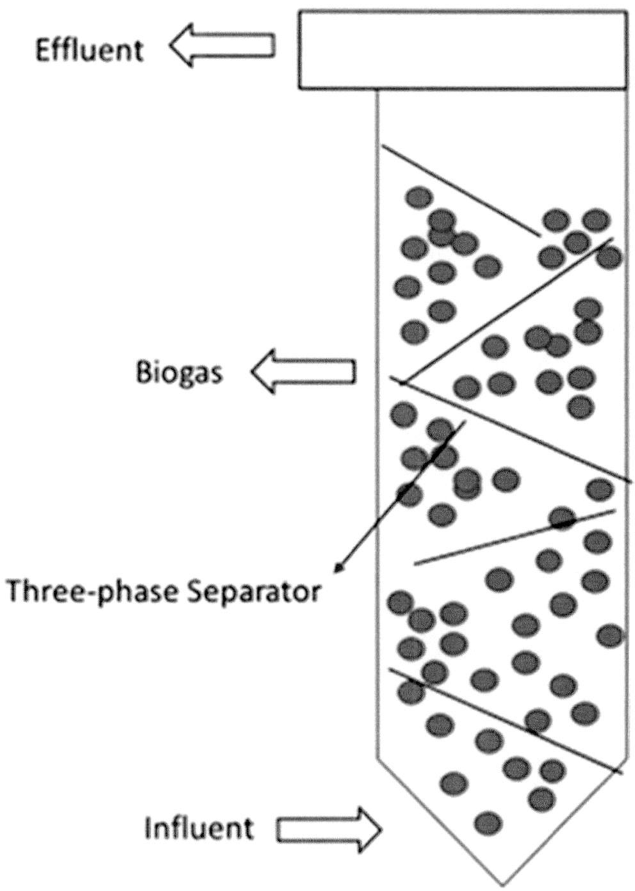

FIGURE 2.5 Schematic three-stage parting baffle.

The internal components operating in the spiral up-flow reactor (SUFR) have been analyzed by Guyuan et al. (2004) and a certain tilt multilayer deflector powered by computer simulation. Honglin et al. (2007) and Yuan et al. (2005) developed the plurality of rotor blades used the used inner longitudinal elements used in a fluidized bed reactor to assist with coagulation, granulation, and biological decay of the reactor.

2.5.9.3 BIOLAYER-PACKAGING MATTER

The first of its type to investigate the internal framework and composition of the biofilm placed activated carbon particles in a fluid bed dosing (Weber et al., 1978). Grainy activated carbon dosage offers several functions, which are discussed here. It results in the enormous growth of specific microbial attachment areas, increased biosorption and biodegradation of the synergies, significant improvement in the matrix of particles and the concentration of oxygen, effective improvement of microbial oxidation efficiency, effective relief of hydraulic scrap, and high levels of microorganisms. McCarty and Smith (1986) introduced fixed filler as an internal portion in the bioreactor in 1986. Strong anaerobic active sludge concentrations were reached, and an anaerobic filter (AF) was created (Ping, 2006). Several researchers have subsequently incorporated activated carbons (Tanyolac and Beyenal, 1998), sand (Rabah and Dahab, 2004), calcium alginate (Tam and Wong, 2000), and porous polymer carrier (Chen et al., 2005) as an inner component in AF, and achieved some remarkable efficiency.

2.5.10 OPERATING TESTS IN ANAEROBIC DIGESTERS

2.5.10.1 IMPORTANCE OF OPERATIONAL CONTROL

The benefits of any wastewater treatment method can only be achieved in an automated way because a structured sequence of actions is performed if it is aerobic or anaerobic (Figure 2.6).

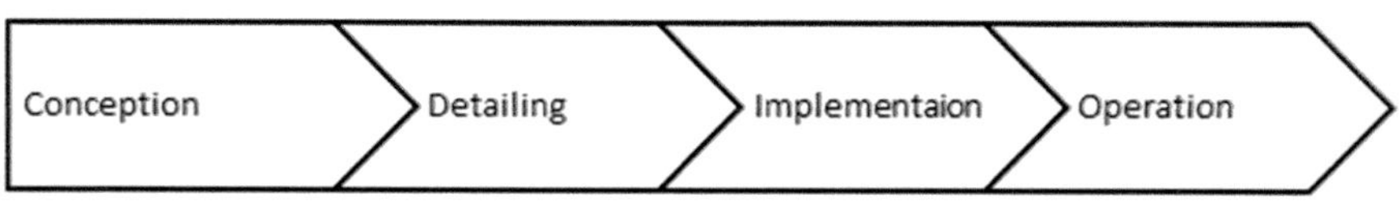

FIGURE 2.6 Function model for a municipal management program.

From the aforementioned flowchart, it is concluded that the key objectives of a wastewater treatment facility, i.e., public health security and ecosystem conservation, will only be achieved if the treatment system is properly planned, well-constructed, well-installed, and often properly monitored. Usually, new treatment plants are constructed in countries with no expertise in handling wastewater, based on standards that are not inherently reliable and manufactured from foreign sources, several times. Those parameters will also be tested during the operating period of the device, attracting into justification the standards initially presumed throughout the design part. The numerous important parameters to be tested during the operational phase of the machine include:

- Influential flow speeds.
- Prominent wastewater physicochemical and microbiological characteristics.
- Preliminary care devices efficiency and operational issues.
- Construction and features of the substance contained in the panels and the chamber gravel.
- The competence of anaerobic reactors and operating problems.
- The volume and features of the biogas created in the anaerobic tank.

2.5.10.2 TREATMENT NETWORK OPERATING POWER

While one of the key points is the organizational flexibility of anaerobic treatment systems, the availability of operating and maintenance staff is a required requirement for ensuring adequate efficiency. The three main monitoring tasks of the care program are:

- *Activity:* refers to the routine or seasonal tasks required to ensure the consistent and reliable output of the conduct facility.
- *Maintenance:* applies to the operations that maintain the facilities of the treatment facility in good condition.
- *Information:* applies to correspondence, typically in writing, with the various parties involved, while at the same time keeping a record of the operation.

2.5.10.3 START TIME OF ANAEROBIC DIGESTERS

Reducing the start time and optimizing anaerobic process operational control are essential factors in rising the performance and productivity of high-rate anaerobic systems. However, it is challenging to examine more thoroughly the parallels, disparities, and compensations of the dissimilar anaerobic systems which are of high rate concerning start time, activity, and control, since the plant conduct becomes essentially based on the properties of the wastewater to be handled.

The start time and activity of anaerobic tanks on a smaller scale are found by technicians to be a challenge, possibly due to poor familiarity with the use of inadequate operational techniques. Therefore, systematized operating measures are very significant throughout the initialization of systems with a high rate, mostly in UASB vessels. The initial transient process describes the start time of anaerobic vessels, defined by operational instabilities. Using seed sludge appropriate for handling wastewater: the system is easily and satisfactorily installed because there is no sludge acclimatization need. Using seed sludge that is not appropriate for handling wastewater. The procedure is implemented through a microbial selection phase after an acclimatization period in this circumstance. Without the use of inoculum is called the utmost dangerous solution to beginning, the process since the reactor needs to be inoculated in the strong wastewater with its microorganisms discovered. Provided the population of microorganisms in wastewater is very less, the time required to endure and accrue a bulky microbial film for 3-7 months.

2.5.10.4 OPERATIONAL TROUBLESHOOTING

Based on the research of Chernicharo et al. (1999), the following features provide a collection of details that can help identify and address operating issues in anaerobic reactors (Table 2.4).

2.6 CONCLUSIONS

Anaerobic treatment is an established and effective process for generating biogas (methane) that can be used to generate green heat and electricity, as well as a compost such as production. Three major characteristics of the anaerobic bioreactor (AB) are inhomogenous framework, period instability, and space instability. Anaerobic treatment performance has an in-depth impact with multiple parameters, such as temperature, pH, OLR, SRT, HRT, up-flow speed and sizes, enabling anaerobic reactor to be

TABLE 2.4 Drift and Features of the Influent

Examination	Plausible basis	Validate	Explanation
Flow constantly less than the expected value	Population or per capita contribution lower than the design value	Flow measuring device	Increase served population
Flow suddenly lower than the expected one	Blockages in sewerage system	Overflow in the contribution area	Unblock sewers
Flow always higher than the expected one	Population or per capita contribution higher than the design value	Flow measuring device	Increase treatment capacity
Daily peaks higher than the expected ones	Equalization lower than the expected one	Flow measuring device	Consider equalization tank
Sudden irregular peaks	Combined system or cross-connection with stormwater sewers	Coincidence with rains	Disconnect illegal connections
Flow sometimes higher than the expected one	Huge mixing of groundwater	Coincidence with rains	Find the infiltration points
pH higher or lower than normal	Industrial wastewater	Presence of illicit sources	Find and take action
Temperature higher or lower than the normal	Industrial waste	Presence of illicit sources	Find and take action
Settleable solids larger than normal	Illicit discarding of local or commercial solid wastes in sewerage,	Type of the settleable solids	Find and take action

operating in a smooth (gas-liquid-built) environment. Therefore, anaerobic treatment involves a specific form of climate, as successful anaerobic processes depend on live bacteria and reactor growth. Inner components of the reactor play an essential role in enhancing the efficiency of the cycle with enhancing efficiently the retention sludge reactor capacity and increasing dramatically the quality of fluidization, contentious bubbles and increased mass transmission. The central transverse materials, the longitudinal materials and the biofilm packing content may be isolated.

KEYWORDS

- **anaerobic digestion**
- **chemical oxygen demand**
- **digester characteristics**
- **extended granular sludge bed**
- **hydraulic retention time**
- **internal circulation**
- **process parameters**
- **solids retention time**
- **up-flow anaerobic sludge blanket**

REFERENCES

Abbasi, T., Tauseef, S. M., & Abbasi, S. A., (2012). Anaerobic digestion for global warming control and energy generation: An overview. *Renew. Sust. Energy Rev., 16*, 3228–3242.

Agler, M. T., Aydinkaya, Z., Cummings, T. A., Beers, A. R., & Angenent, L. T., (2010). Anaerobic digestion of brewery primary sludge to enhance bioenergy generation: A comparison between low- and high-rate solids treatment and different temperatures. *Bioresour. Technol., 101*, 5842–5851.

Aiyuk, S., Forrez, I., Lieven, D. K., Van, H. A., & Verstraete, W., (2006). Anaerobic and complementary treatment of domestic sewage in regions with hot climates: A review. *Bioresour. Technol.*, 2225–2241.

Berger, S., Welte, C., & Deppenmeier, U., (2012). Acetate activation in *Methanosaeta thermophila*: Characterization of the key enzymes pyrophosphatase and acetyl-CoA synthetase. *Archaea*.

Buijs, C., Heertjes, P. M., & Van, D. M. R. R., (1982). Distribution and behavior of sludge in upflow reactors for anaerobic treatment of wastewater. *Biotechnol. Bioeng., 24*, 1975–1989.

Chelliapan, S., Wilby, T., & Sallis, P. J., (2006). Performance of an up-flow anaerobic stage reactor (UASR) in the treatment of pharmaceutical wastewater containing macrolide antibiotics. *Water Res., 40*, 507–516.

Chen, J., Tang, C., Zheng, P., & Zhang, L., (2008). Performance of lab-scale SPAC anaerobic bioreactor with high loading rate. *Chinese J. Biotechnol., 24*, 1413–1419.

Chen, X. G., Dayong, Y., & Zhaoji, H., (2005). Experimental study on a three-phase outer circulating fluidized bed by a sequencing batch reactor of biofilm process. *Seminar on Engineering Education Cooperation & Academic Research for Chinese-French Universities*, 367370.

Chen, X. G., Zheng, P., Cai, J., & Qaisar, M., (2010). Bed expansion behavior and sensitivity analysis for super-high-rate anaerobic bioreactor. *J. Zhejiang University, 11*, 79– 86.

Chen, X. G., Zheng, P., Guo, Y. J., Mahmood, Q., Tang, C. J., & Ding, S., (2010). Flow patterns of super-high-rate anaerobic bioreactor. *Bioresour. Technol., 101*, 7731–7735.

Chen, X. G., Zheng, P., Qaisar, M., & Tang, C. J., (2012). Dynamic behavior and concentration distribution of granular sludge in a super-high-rate spiral anaerobic bioreactor. *Bioresour. Technol., 111*, 134–140.

Chen, X., Ping, Z., Ding, S., et al., (2011). Specific energy dissipation rate for super-high-rate anaerobic bioreactor. *J. Chem. Technol. Biotechnol., 86*, 749–756.

Chernicharo, C. A. L., & Campos, C. M. M., (1995). *Anaerobic Sewage Treatment* (p. 65). Department of Sanitary and Environmental Engineering at UFMG School of Engineering. Apostila.

Chernicharo, C. A. L., Van, H. A. C., & Cavalcanti, P. F. F., (1999). Chapter 9: Operational control of anaerobic reactors. *Sanitary Sewage Treatment by Anaerobic Process and Controlled Disposal in the Soil* (p. 436). Rio de Janeiro, Brazil.

Chernicharo, C. A. L., Van, H. A. C., Cybis, L. F., & Foresti, E., (2001). Post-treatment of anaerobic effluents in Brazil: State-of-the-art. *Proceedings of the 9th World Congress on Anaerobic Digestion* (pp. 747–752). Technologisch Instituut, IWA, Netherlands Association for Water Management. Antwerp, Belgium.

Chong, S., Sen, T. K., Kayaalp, A., & Ang, H. M., (2012). The performance enhancements of upflow anaerobic sludge blanket (UASB) reactors for domestic sludge treatment – a state-of-the-art review. *Water Res., 46*, 3434–3470.

El-Mashad, H. M., Zeeman, G., van Loon, W. K. P., Bot, G. P. A., & Lettinga, G., (2004). Effect of temperature and temperature fluctuation on thermophilic anaerobic digestion of cattle manure. *Bioresour. Technol., 95*, 91–201.

Elmitwalli, T. A., Zandvoort, M. H., Zeeman, G., Burning, H., & Lettinga, G., (1999). Low temperature treatment of domestic sewage in upflow anaerobic sludge blanket and anaerobic hybrid reactors. *Water Sci. Technol., 39*, 177–185.

Fang, H. H. P., & Jia, X. S., (1998). Soluble microbial products (SMP) of acetotrophic methanogenesis. *Bioresour. Technol., 66*, 235–239.

Florencio, L., Nozhevnikova, A., Van, L. A., Stams, A. J. M., Field, J. A., & Lettinga, G., (1993). Acidophilic degradation of methanol by a methanogenic enrichment culture. *FEMS Microbiol. Lett., 109*, 1–6.

Friedman, A. A., & Tait, S. J., (1980). Anaerobic rotating biological contactor for carbonaceous wastewater. *J. Water Pollut. Contr. Fed.*, 2257–2269.

Gao, W. J., Leung, K. T., Qin, W. S., & Liao, B. Q., (2011). Effects of temperature and temperature shock on the performance and microbial community structure of a submerged anaerobic membrane bioreactor. *Bioresour. Technol., 102*, 8733–8740.

Ge, H., Jensen, P. D., & Batstone, D. J., (2011). Increased temperature in the thermophilic stage in temperature phased anaerobic digestion (TPAD) improves degradability of waste activated sludge. *J. Hazard. Mater., 187*, 355–361.

Guyuan, L., Junfeng, D., Fangying, J. I., & Ning, L., (2004). Study on the characteristic of spiral up-flow reactor system and its performance on biological nitrogen and phosphorus removal. *Acta Scientiae Circumstantiate, 24*, 15–20.

Halalsheh, M., Koppes, J., Den, E. J., Zeeman, G., Fayyad, M., & Lettinga, G., (2005). Effect of SRT and temperature on biological conversions and the related scum forming potential. *Water Res., 39*, 2475–2482.

Hansen, K. H., Angelidaki, I., & Ahring, B. K., (1998). Anaerobic digestion of swine manure: Inhibition by ammonia. *Water Res., 32*, 5–12.

Honglin, Y., Yongjun, L., & Xiaochang, W., (2007). Microbial community structure and dynamic changes of fluidized-pellet-bee bioreactor. *J. Microbiol., 27*, 1–5.

Hulshoff, P. L. W., Lopes, S. I. D. C., Lettinga, G., & Lens, P. N. L., (2004). Anaerobic sludge granulation. *Water Res., 38*, 1376–1389.

Jewell, W. J., (1981). Development of the attached microbial film expanded bed process for aerobic and anaerobic waste treatment. In: Cooper, P. F., & Atkinson, (eds.), *Biological Fluidized Bed Treatment of Water and Wastewater* (p. 251). Ellis Horwood, Chichester.

Jewell, W. J., (1987). Anaerobic sewage treatment. *Environ. Sci. Technol., 21*, 14–21.

Ji, T., Luo, G., Wang, D., Xu, X., & Zhu, L., (2007). Effect of prolonged sludge age on biological nutrient removal in spiral up-flow reactor system and flow pattern interpretation. *J. Chem. Ind. Eng., 58*, 2613–2618.

Kalyuzhnyi, S., Estrada, D. L. S. L., & Martinez, J. R., (1998). Anaerobic treatment of raw and pre clarified potato-maize wastewaters in a UASB reactor. *Bioresour. Technol., 66*, 195–199.

Kangle, K. M., Kore, V. S., & Kulkarni, G. S., (2012). Recent trends in anaerobic codigestion: A review. *Univ. J. Environ. Res. Technol., 2*, 210–219.

Karim, K., Thoma, G. J., & Al-Dahhan, M. H., (2007). Gas-lift digester configuration effects on mixing effectiveness. *Water Res., 41*, 3051–3060.

Katarzyna, B. D. C., & Flaherty, V. O., (2013). Low-temperature (10°C) anaerobic digestion of dilute dairy wastewater in an EGSB bioreactor: Microbial community structure, population dynamics, and kinetics of methanogenic populations. *Archaea*.

Kato, M. T., Field, J. A., Versteeg, P., & Lettinga, G., (1994). Feasibility of expanded granular sludge bed reactors for the anaerobic treatment of low-strength soluble wastewaters. *Biotechnol. Bioeng., 44*, 469–479.

Kayhanian, M., (1995). Biodegradability of the organic fraction of municipal solid waste in a high-solids anaerobic digester. *Waste Manage. Res., 13*, 123–136.

Kayhanian, M., (1999). Ammonia inhibition in high-solids biogasification: An overview and practical solutions. *Environ. Technol., 20*, 355–365.

Lens, P. N. L., Van, D. B. M. C., Hulshoff, P. L. W., & Lettinga, G., (1998). Effect of staging on volatile fatty acid degradation in a sulfidogenic granular sludge reactor. *Water Res., 32*, 1178–1192.

Lettinga, G., (1995). *Anaerobic Reactor Technology: Reactor and Process Design*. International Course on Anaerobic Treatment. Wageningen Agricultural University/IHE Delft, Wageningen.

Lettinga, G., (1995). *Introduction in International Course on Anaerobic Treatment* (pp. 17–28) Wageningen Agricultural University/IHE Delft, Wageningen.

Lettinga, G., Field, J. A., Sierra-Alvarez, R., Van, L. J. B., & Rintala, J., (1991). Future perspectives for the anaerobic treatment of forest industry wastewaters. *Water Sci. Technol., 36*, 91–102.

Lettinga, G., Rebac, S., & Zeeman, G., (2001). Challenge of psychrophilic anaerobic wastewater treatment. *Trends Biotechnol.*, 363–370.

Lettinga, G., Van, V. A. F. M., Hobma, S. W., De Zeeuw, W., & Klapwijk, A., (1980). Use of the upflow sludge blanket (USB) reactor concept for biological wastewater treatment, especially for anaerobic treatment. *Biotechnol. Bioeng., 22*, 699–734.

Madigan, M. T., Martinko, J. M., Stahl, D., & Clark, D. P., (2008). *Brock Biology of Microorganisms*. Benjamin-cummings, reading, mass, USA.

Mah, R. A., Smith, M. R., & Baresi, L., (1978). Studies on an acetate-fermenting strain of *Methanosarcina*. *Appl. Environ. Microbiol., 35*, 1174–1184.

Mahmoud, N., Zeeman, G., Gijzen, H., & Lettinga, G., (2003). Solids removal in upflow anaerobic reactors: A review. *Bioresour. Technol., 90*, 1–9.

Mahmoud, N., Zeeman, G., Gijzen, H., & Lettinga, G., (2004). Anaerobic stabilization and conversion of biopolymers in primary sludge – effect of temperature and sludge retention time. *Water Res., 35*, 983–991.

McCarty, P. L., & Smith, D. P., (1986). Anaerobic wastewater treatment. *Environ. Sci. Technol., 20*, 1200–1206.

McCarty, P. L., (1964). *Anaerobic Waste Treatment Fundamentals* (pp. 107–112). Public Works.

Meynell, P. J., (1976). *Methane: Planning a Digester*. Schocken Books, New York, NY, USA.

Mittal, G. S., (2006). Treatment of wastewater from abattoirs before land application—a review. *Bioresour. Technol., 97*, 1119–1135.

Nanda, S., & Berruti, F., (2021). Municipal solid waste management and landfilling technologies: A review. *Environ. Chem. Lett*, 19, 1433–1456.

Ostrem, K., (2004). *Greening Waste: Anaerobic Digestion for Treating the Organic Fraction of Municipal Solid Wastes*. Earth Engineering Center Columbia University.

Rabah, F. K. J., & Dahab, M. F., (2004). Biofilm and biomass characteristics in high-performance fluidized-bed biofilm reactors. *Water Res., 38*, 4262–4270.

Reddy, P. J., (2011). *Municipal Solid Waste Management: Processing- Energy Recovery-Global Examples.* CRC Press.

Ruiz, I., Veiga, M. C., De, S. P., & Blázquez, R., (1997). Treatment of slaughterhouse wastewater in a UASB reactor and an anaerobic filter. *Bioresour. Technol., 60*, 251–258.

Seghezzo, L., Zeeman, G., Van, L. J. B., Hamelers, H. V. M., & Lettinga, G., (1998). A review: The anaerobic treatment of sewage in UASB and EGSB reactors. *Bioresour. Technol.*, 175–190.

Stronach, S. M., Rudd, T., & Lester, J. N., (1986). *Anaerobic Digestion Processes in Industrial Wastewater Treatment*. Springer-Verlag, Berlin.

Sung, S., & Liu, T., (2003). Ammonia inhibition on thermophilic anaerobic digestion. *Chemosphere, 53*, 43–52.

Takashima, M., & Speece, R. E., (1989). Mineral nutrient requirements for high-rate methane fermentation of acetate at low SRT research. *J. Water Pollut. Contr. Fed., 61*, 1645–1650.

Tam, N. F. Y., & Wong, Y. S., (2000). Effect of immobilized microalgal bead concentrations on wastewater nutrient removal. *Environ. Pollut., 107*, 145–151.

Tanyolac, A., & Beyenal, H., (1998). Prediction of substrate consumption rate, average biofilm density and active thickness for a thin spherical biofilm at pseudo steady state. *Biochem. Eng. J., 2*, 207–216.

Van, L. J. B., Sanz, M. J. L., & Lettinga, G., (1996). Effect of temperature on the anaerobic thermophilic conversion of volatile fatty acids by dispersed and granular sludge. *Water Res., 30*, 199–207.

Varel, V. H., Hashimoto, A. G., & Chen, Y. R., (1980). Effect of temperature and retention time on methane production from beef cattle waste. *Appl. Environ. Microbiol., 40*, 217–222.

Von, S. M., (1995). *Introduction to Water Quality and Sewage Treatment* (p. 240). Department of Sanitary and Environmental Engineering – UFMG, Belo Horizonte.

Wang, C. H., Zheng, P., Chen, J. W., Tang, C. J., & Yu, Y., (2009). Kinetics analysis of spiral automatic circulation anaerobic reactor. *J. Zhejiang U., 35*, 222–227.

Weber, W. J. Jr., Pirbazari, M., & Melson, G. L., (1978). Biological growth on activated carbon: An investigation by scanning electron microscopy. *Environ. Sci. Technol., 12*, 817–819.

Wei, C., Li, L., Wum, J., Wu, C., & Wu, J., (2007). Influence of funnel-shaped internals on hydrodynamics and mass transfer in internal loop three-phase fluidized bed. *J. Chem. Ind. Eng., 58*, 591–595.

Wijetunga, S., Li, X. F., & Jian, C., (2010). Effect of organic load on decolorization of textile wastewater containing acid dyes in upflow anaerobic sludge blanket reactor. *J. Hazard. Mater., 177*, 792–798.

Winkler, M. K., Bassin, J. P., Kleerebezem, R., Van, D. L. R. G., & Van, L. M. C., (2012). Temperature and salt effects on settling velocity in granular sludge technology. *Water Res., 46*, 3897–3902.

Yanling, H., (1998). *Anaerobic Biological Treatment of Wastewater*. China Light Industry Press, Beijing, China.

Yuan, H. L., & Liu, Y. J., (2005). Pilot study of a fluidized pellet—bed bioreactor for simultaneous biodegradation and solid/liquid separation in municipal wastewater treatment future of urban wastewater systems decentralization and reuse. *Proc. IWA Conf.*, 253–260.

Zeeman, G., Sanders, W. T. M., Wang, K. Y., & Lettinga, G., (1997). Anaerobic treatment of complex wastewater and waste activated sludge-application of an upflow anaerobic solid removal (UASR) reactor for the removal and prehydrolysis of suspended COD. *Water Sci. Technol., 35*, 121–128.

Zheng, K., Ji. J. Y., Xing, Y. J., & Zheng, P., (2012). Hydraulic characteristics and their effects on working performance of compartmentalized anaerobic reactor. *Bioresour. Technol., 116*, 47–52.

Zheng, P. F. X., (2006). *Waste Biological Treatment*. Higher Education Press, Beijing, China.

CHAPTER 3

Metabolic Engineering of Methanogenic Archaea for Biomethane Production from Renewable Biomass

RAJESH KANNA GOPAL,[1] PREETHY P. RAJ,[2] AJINATH DUKARE,[3] and ROSHAN KUMAR[4]

[1]*Department of Plant Biology and Plant Biotechnology, Presidency College, Chennai, Tamil Nadu, India*

[2]*Department of Biotechnology, University of Madras, Chennai, Tamil Nadu, India*

[3]*Chemical and Biochemical Processing Division, Indian Council of Agricultural Research - Central Institute for Research on Cotton Technology, Mumbai, Maharashtra, India*

[4]*Department of Human Genetics and Molecular Medicine, Central University of Punjab, Bathinda, Punjab, India*
E-mail: roshan.kumar@cup.edu.in (Roshan Kumar)

ABSTRACT

The methanogenic archaea are biomethane-producing microorganisms that perform anaerobic digestion of organic wastes. Anaerobic digestion is a closed process with an array of biochemical pathways for the production of biogas. However, this process involves different kinds of microbiota in every step for industrial-scale wastewater treatment. It is a promising technology, not only for biowaste treatment but also for the generation of biomethane (bioenergy production). In recent decades, advanced technologies and reactors were developed for efficient biomethane production. However, the complete utilization of bioorganic waste depends on the bioaugmentation and metabolic engineering technologies. Thus, genetic engineering deals with the alteration of the biochemical pathway in methanogenic archaea to short-cut unwanted steps or the addition of pathway to yield enhanced biogas. Hence, this chapter deals with the discussion of advanced genetic engineering tools and techniques implemented to enhance biomethane production.

3.1 INTRODUCTION

It is essential to develop advanced technology for zero waste treatment methods through an all-inclusive waste management system and biofuel production (Nanda and Berruti, 2021a,b,c). Anaerobic digestion or fermentation is an efficient process in the digestion of organic wastes and energy production by the generation of biogas (biomethane). Biomethane is one of the biofuels for energy production from waste management, which is possible through potential methanogenic bacteria (Holm-Nielsen et al., 2009; Weiland, 2010). The by-products of anaerobic digestion are biomethane (CH_4) (storable energy carrier), CO_2 and other residual matter (used as biofertilizer) (Risberg et al., 2017).

The methanogenic bacteria are the potential catalyzers for the production of biomethane, which are categorized under diverse archaeal groups (Balch et al., 1979). Biomethane is highly flammable and is considered alternate energy for fossil-based energy consumption in the upcoming decades (Ren et al., 2008). Methanogenesis occur naturally in the gut of animals, oil fields, swamps, and biomass-degrading ecosystem (Garcia et al., 2000) and are implemented in wastewater treatment plants integrated with biogas production. Agricultural biomass waste, municipal solid waste, food waste, and other food industrial wastes can also be used as a feedstock for the production of biofuel (bioethanol and biomethane) (Nanda et al., 2014, 2015; Srivastava, 2019; Okolie et al., 2021) (Figure 3.1). The residual biomass wastes from seaweed-based (macroalgae) chemical industries also serves as a suitable feedstock for the production of biomethane. During recent decades, metabolic stimulants were added to accelerate the rate of reaction and enhanced biogas production (Xia et al., 2016; Patterson et al., 2011).

3.2 BIOMETHANE PRODUCTION

The anaerobic digestion of organic matter involves four stages process. In the first stage, hydrolysis of complex organic macromolecules such as proteins, polysaccharides, and lipids takes place by the action of extracellular enzymes into amino acids, sugars, long-chain fatty acids, respectively (Vavilin et al., 2008). Acidogenesis (fermentation) is the second stage, where oxidation of the end products from the first stage takes place to synthesize products such as acetate, propionate, butyrate, formate, ethanol, H_2 and CO_2. The third stage is acetogenesis, in which all the end products of the second stage are further oxidized to acetate and CO_2 with H_2 formation (Batstone et al., 2002). In this exergonic reaction, a high amount of energy is generated for the organisms by keeping very low H_2 pressure (McInerney et al., 2008). Conversion of acetate, other methylated compounds, CO_2 and H_2 into biomethane is the final stage, which is carried out only by the methanogenic archaea. Therefore, it is very clear that the consortium of microbes is a successful recipe for biogas production, which may vary between different kinds of reactors (Solli et al., 2014). During anaerobic digestion in a biogas reactor, injection of H_2 enhances biomethane production by methanogenic archaea (Kougias et al., 2017).

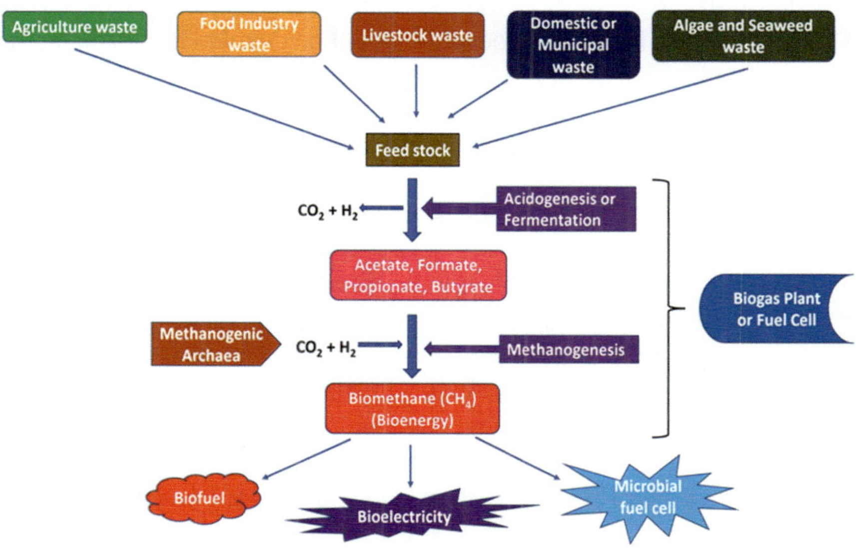

FIGURE 3.1 Biowaste management and biomethane production through anaerobic degradation.

The methanogenic archaea use H_2 and CO_2 along with formate, acetate, and methylamine to generate biomethane (Enzmann et al., 2018). The biogas composed of CH_4 (50–70%), and CO_2 (30–40%) with trace amounts of hydrogen sulfide (H_2S) (Srivastava, 2019; Awe et al., 2017). The composition of biogas may vary due to different sources of feedstock used by the methanogenic archaea, such as organic waste, landfill sources, and sewage digester (Awe et al., 2017). Biomethane and biohydrogen are used as stock material for the production of useful biochemicals such as single-cell proteins, biopolymers, lipids, etc. (Strong et al., 2015).

Biomethane combustion with oxygen releases energy that can be used for cooking and vehicle fuel. Anaerobic digestion of wastewater is effective in the generation of electrons in biofuel cells for bioelectricity production. However, methanogenic metabolism, optimization of conditions and synchronizing of new technologies are to be upgraded for biomass conversion into biomethane in biogas production plants (Enzmann et al., 2018). Different kinds of bioreactors were designed according to their respective feedstocks (substrate) for biomethane gas production. It involves different kinds of microorganisms depending on their community profile and growth parameters (Schnurer, 2016). CO_2 and H_2 are supplemented in addition to the feedstock to enhance the yield and purity of biogas in industrial biogas plants (Enzmann et al., 2018). The utilization of biogas as bioenergy would be doubled in the upcoming decades from 14.5 GW in 2012 to 29.5 GW in 2022 (Raboni and Urbini, 2014; Karaszova et al., 2015).

3.3 METHANOGENIC ARCHAEA

Methanosarcinales is probably the only order in the classification of methanogenic archaea bacteria to generate biomethane from acetate. Methylotrophic and methanogenic archaea bacteria using methylated thiols or methylamines and methanol for the production of biomethane are *Methanomassiliicoccales* and *Methanobacteriales* (Enzmann et al., 2018). Vanwonterghem et al. (2016) and Evans et al. (2015) and reported that two new phyla *Verstraetearchaeaota* and *Bathyarchaeota* as methylotrophic bacteria undergoes methanogenesis. Methanogens belong to anoxic habitats dispersed only in extreme ecosystems include hydrothermal vents or high saline lakes (Garcia et al., 2000). *Methanopyrus kandleri* and *Methanocaldococcus jannaschii* were isolated separately from the black smoker and white smoker chimney of the Gulf of California (2000 m depth) and East Pacific Rise (2600 m depth), respectively (Kurr et al., 1991; Jones et al., 1983).

Asakawa et al. (1995) have isolated many different strains of *Methanobacterium* such as *Methanobrevibacter arboriphilus* and *Methanosarcina mazei* TMA from the rice fields. Brauer et al. (2011) isolated *Methanoregula boonei* from acidic peat bog. Mathrani et al. (1988) isolated and cultured halophilic methanogen *Methanohalophilus zhilina* from the saline lake in Egypt. An interesting factor is that *Methanobacterium arbophilicum* uses hydrogen from the decaying wet-wood tissue through the degradation of cellulose and pectin by *Clostridium butyricum* for biomethane production (Zeikus and Henning, 1975; Schink et al., 1981). In human feces, *Methanomassiliicoccus luminyensis*, *Methanosphaera stadtmanae*, and *Methanobrevibacter smithii* have been detected (Miller et al., 1982; Dridi et al., 2012). *Methanobacterium oralis* (Robichaux et al., 2003), *Methanosphaera* sp. (Ferrari et al., 1994) and *Methanosarcina* sp. (Belay et al., 1988) have been isolated from the dental plaque of humans. Some of the multicellular, methanogenic bacteria are reported in the genus *Methanosarcina*, *Methanobacterium*, and *Methanolobus* (Kern et al., 2015, 2016; Mochimaru et al., 2009).

3.4 BIOAUGMENTATION

Hydrolysis and methanogenesis are the two major steps in biogas production. Bioaugmentation refers to enhancing the biochemical reactions of these steps for high yield of biogas. For lignocellulosic degradation of biogas production, bioaugmentation with cellulose-degrading microbes, hydrolytic enzymes, and anaerobic fungi are auspicious in enhanced synthesis of biomethane production (Oner et al., 2018; Speda et al., 2017; Peng et al., 2014; Zhang et al., 2015; Nanda et al., 2017). Some of them are *Clostridium cellulolyticum*, *Caldicellulosiruptor lactoaceticus,* and *Acetobacteroides hydrogenigenes.* Thus, a consortium of different cellulolytic microbes (Nzila, 2017) and high endoglucanase activity is effective in biogas yield in maize silage (Poszytek et al., 2016).

Supplementation of hydrolytic enzymes for cellulose and xylan degradation is effective in anaerobic digestion (Azman et al., 2017). Through bioaugmentation, *Coprothermobacter proteolyticus* has been identified for the hydrolysis of proteinaceous feedstock followed by fermentation (Sasaki et al., 2011). Furthermore, in the case of fat as feedstock, lipases or *Clostridium lundense* and *Syntrophomonas zehnderi* could be effective (Nzila, 2017; Cirne et al., 2007). However, in most of the studies, it affected the anaerobic digestion process of methanogenic bacteria (*Methanobacterium formicicum*) (Silva et al., 2014) by generating long-chain fatty acids and ammonia, which highly inhibits the fermentation process (Wang et al., 2017). *Methanoculleus bourgensis* is resistant to a high amount of ammonia, which could be effective in anaerobic digestion along with other coculturing microbes (Karakashev *et al.*, 2006; Fotidis et al., 2014). By bioaugmentation, *Clostridium cellulovorans*, *Pseudobutyrivibrio xylanivorans* Mz5T and *Fibrobacter succinogenes* S85 were found effective in the treatment of brewery grain waste for biomethane production (Cater et al., 2015).

3.5 METABOLIC ENGINEERING

The commercial biomethane production system completely relies on the utilization and development of metabolic pathway engineering techniques and genetic modifications. The metabolic pathway is nothing, but a series of biochemical reactions, which are optimized via the application of genetic tools to enhance the rate of biochemical reactions to yield high content of biofuel (Beer et al., 2009; Ho et al., 2013; Lee et al., 2013; Mukhopadhyay, 2015; Papilo et al., 2017). The suitable and potential archaeal strain is employed and its growth factors are optimized for industrial-scale production of bioenergy (Lee et al., 2013). For example, optimization of cheap substrate concentration to attain the maximum growth rate of *Saccharomyces cerevisiae* strain CICC 1308 was studied for bioethanol production (Jin et al., 2012).

Inoculum size of microbes can be optimized to degrade wheat and rice straw in batch cultivation for wastewater treatment and biogas plants. *Clostridium cellulolyticum* is efficient for high degradation of rice and wheat straw and identified based on molecular characterization using random fragments of the *cel5* community (Sun et al., 2016). Variations in the microbial community structure were observed during biogas production due to high salinity stress, and which was identified by Illumina high-throughput sequencing technology. Therefore, hydrogen using methanogenic archaea are found to be lower in resistance against high salinity concentrations that acidogenic methanogens (Wang et al., 2017).

Biogas generating microbial communities have a characteristic feature of hydrogen metabolism, which is the key feature for optimal biogas production. These features are characterized and identified by novel high-throughput metagenomic analysis among the microbial communities for their taxonomical complexity of consortia systems to attain a high yield of biogas (Jaenickie et al., 2011). The biomethane production involves two different microbial communities. One group named acidogenesis (*Clostridia* spp.) performs decomposition of organic matter and generates acetate, formate, other methyl-by-products, and hydrogen. The other microbial community uses the by-products of acidogenic bacteria and undergoes methanogenesis *Methanoculleus marisnigri* (Wirth et al., 2012). To enhance or stimulate the anaerobic digestion process, genetic manipulation of methanogenic bacteria is essential (Caruso et al., 2019).

In biomethane production, CO_2 is first reduced to form activated formyl-methanofuran (Wagner et al., 2016) by reduced ferredoxin as an electron donor. Next, the formyl group is transferred to tetrahydromethanopterin in the second reaction. Dehydration and reduction take place in the formyl group to form methylene-tetrahydromethanopterin with reduced F420 as an electron donor (Liu and Whitman, 2008). Coenzyme M acts as a transferase and transfers methyl group from the methylene-tetrahydromethanopterin. Lastly, coenzyme B (CoB) as an electron donor reduces methyl-CoM to generate methane. Then the reduction of residual heterodisulfide (CoM-S-S-CoB) took place by the action of H_2 to recycle coenzymes (Thauer et al., 2008; Liu and Whitman, 2008). In contrast to this, *Methanothermobacter thermautotrophicus* and *Methanosarcina barkeri* oxidize four molecules of CO to form CO_2 by CO dehydrogenase enzyme and subjected to reduce into one molecule of CO_2 to synthesize CH_4 with H_2 molecule as an electron donor (O'Brien et al., 1984; Daniels et al., 1977).

3.5.1 INDUCED MUTAGENESIS, CONJUGATION, AND POLYETHYLENE GLYCOL-MEDIATED TRANSFORMATION

Mutagenesis and conjugation are the most common and widely implemented methods for genetic manipulation in bacteria. In this regard, the induced mutation of wild strains may result in muted microorganisms with the desired capacity of biomass-degrading ability to synthesize biogas in a short time. Several methanogenic archaeal strains are mutated for their inhibitor resistance to enhance biogas production. As a result, the biodegrading activity of the mutant methanogenic archaea includes *Methanococcus maripaludis*, *Methanococcus voltae* PS. and *Methanosarcina* and transformed strains of *Escherichia coli*, *Lactobacillus*, *Klebsiella*, and

Clostridium sp. were enhanced in comparison with the wild strains (Rother and Metcalf, 2005; Senthilkumar and Gunasekaran, 2005). Thus, genetic manipulations are effective for targeted engineering to remove unwanted biochemical reactions or the addition of useful biochemical reactions for enhanced production of biofuels.

The most important methods for genetic manipulation are isolation of clonal populations (plating on solidified media); transfer of genetic material (transformation, conjugation, or transduction) and to link the particular gene of interest to a selected phenotype with marker genes (antibiotics) (Enzmann et al., 2018). Mutated *Methanothermobacter* have been isolated by culturing it on the solid culture media (Hummel and Bock, 1985; Harris and Pinn, 1985). However, the random induced mutation is inefficient in the transfer of the gene of interest in these particular archaea (Worrell *et al*., 1988). In *Methanococcus voltae*, the *pac* gene from *Streptomyces alboniger*, which is resistance to the antibiotic puromycin, has been used as an antibiotic selective marker (Gernhardt et al., 1990). However, in *Methanococcus*, pseudomurein absence from the cell wall, followed by removal of the proteinaceous surface by polyethylene glycol (PEG) facilitates free protoplast, which can be feasible for DNA uptake (Enzmann et al., 2018). Furthermore, the PEG-mediated DNA transformation is ineffective (Oelgeschlager and Rother, 2009). Hence, in *Methanosarcina* spp., the addition of cationic liposomes was reported effective (Metcalf et al., 1997).

3.5.2 GENE INTEGRATION AND DELETION

Among *Methanococcus* species, *M. maripaludis* is effective in consuming H_2 and CO_2 and it is a principal genetic model for hydrogenotrophic methanogens (Sarmiento et al., 2011). In comparison, the gene expression is feasible, when cryptic plasmid vectors, pUBR500 in *Methanococcus* (Tumbula et al., 1997), and pC2A in *Methanosarcina* (Metcalf et al., 1997) could be genetically engineered into shuttle vectors in *E. coli* (fast replicator). Consequently, the unwanted DNA sequence from the chromosome has been deleted by using counterselective markers (Moore and Leigh, 2005; Pritchett et al., 2004). However, direct insertion and deletion of a gene in methanogenic archaea is rather inefficient and relies on the use of gene length between 500 bp and 1000 bp. Therefore, a breakthrough was made in the genetic manipulation of *Methanosarcina*, in which, phage recombination system (C31) of *Streptomyces* had been reported to insert genetic material through site-specific recombination (Guss et al., 2008), and Flp/FRT system of yeast had been reported to remove unwanted genetic material from the chromosome (Welander and Metcalf, 2008). *Methanosarcina* is an attractive genetic model due to its different pathway (routes) for methanogenesis by using carbon monoxide, acetate, methanol, methylamines, and methyl sulfides. The targeted site-specific mutation blocks certain pathways and paves the way for the other instead (Guss et al., 2005; Pritchett and Metcalf, 2005; Welander and Metcalf, 2005).

A fusion of promoter and reporter genes, especially *lacZ* (encodes β-galactosidase from *E. coli*) or *uidA* (encodes β-glucuronidase from *E. coli*) and *hisA* gene (encodes N-[5(-phosphor-ribosyl-formimino]-5-aminol-[5(-phosphoribosyl]4-imidazolecarboxamide isomerase) were implemented for chromosome integration in *Methanococcus* spp. (Beneke et al., 1995; Klein and Horner, 1995). The discovery of the CRISPR/Cas9-system in *Streptococcus pyogenes* (Doudna and Charpentier, 2014) is the most advanced, successful, high-throughput, and versatile applications for gene insertion and deletion in *Methanosarcina acetivorans* (Nayak and Metcalf, 2017). Genetic tools and methodology formulated for one model methano-archaeal organism could be easily useful for another model system, for example, the insect transposable element *Himar1* and *transposase* in *Methanosarcina* (Zhang et al., 2000) could be adapted in *Methanococcus* (Sattler et al., 2013). Figure 3.2 represents the overall schematic representation of metabolic engineering of methanogenic archaea for enhanced biomethane production.

3.6 CONCLUSIONS

Biogas production is a promising technology for food and agriculture waste management and thereby generating renewable energy for a sustainable society. The microbial fuel cell is another option for the generation of bioelectricity via complex organic waste management. It is an alternative energy option for fossil fuel energy, which is going to deplete in the future. Fermentation followed by anaerobic digestion is the biochemical process by consortia of microbes in a microbial system. The improvement and upgradation of biomethane production depend on the potential microbial consortia for the respective substrate, bioaugmentation, optimization, and

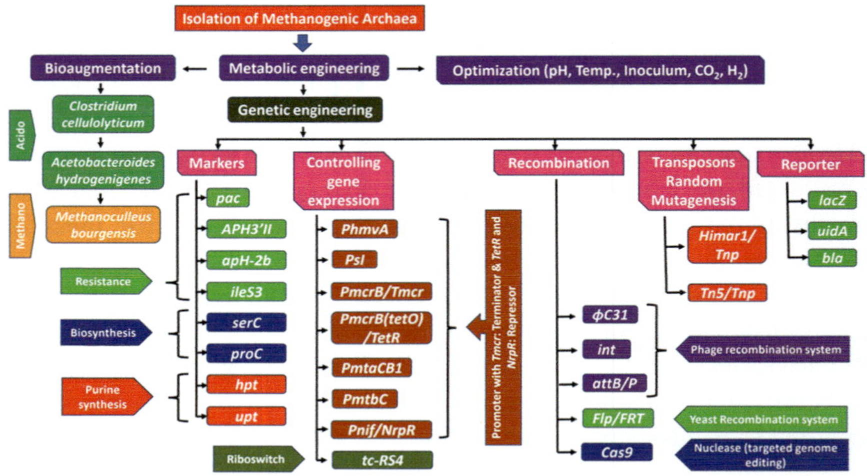

FIGURE 3.2 Overall illustration of metabolic engineering of methanogenic archaea for enhanced biomethane production (Enzmann et al., 2018). Bioaugmentation is fruitful in the selection of potential acidogenic (Acido) microbes and methanogenic archaea (Methano). In genetic manipulation, genes including *pac* (in *Methanococcus* and *Methanosarcina*) (Metcalf et al., 1997), *APH3(II* (in *Methanococcus maripaludis*) (Argyle et al., 1996), *apH-2b* (in *Methanococcus mazei*) (Mondorf et al., 2012), and *ileS3* (in *Methanosarcina acetivorans*) (Bocazzi et al., 2000) were resistance genes, *serC* (*Methanosarcina barkeri*) (Metcalf et al., 1996), and *proC* (in *Methanosarcina acetivorans*) (Pritchett et al., 2004) were biosynthesis genes, and *hpt* (in *Methanosarcina*) (Pritchett et al., 2004), and *upt* (in *Methanococcus maripaludis*) (Moore and Leigh, 2005) were purine synthesis genes altogether used as marker genes; Promoter genes such as *PhmvA* (in *Methanococcus*) (Beneke et al., 1995), *PsI* (*Methanococcus*) (Sun and Klein, 2004), *PmcrB/Tmcr* with terminator gene (in *Methanococcus, Methanosarcina*) (Gernhardt et al., 1990), *PmcrB(tetO)/TetR* with repressor gene (in *Methanosarcina*) (Guss et al., 2008), *PmtaCB1* (in *Methanosarcina acetivorans*) (Rother et al., 2005), *PmtbC* (in *Methanococcus mazei*) (Mondorf et al., 2012), *Pnif/NrpR* with repressor gene (in *Methanococcus maripaludis*) (Lie and Leigh, 2003), *tc-RS4* alone riboswitch (in *Methanosarcina acetivorans*) (Demolli et al., 2014) all are used in controlling gene expression; For genetic recombination, phage recombination system (genes include *ϕC31, int,* and *attB/P* in *Methanosarcina*) (Guss et al., 2008), yeast recombination system (*flp/FRT* in *Methanococcus,* and *Methanosarcina*) (Hohn et al., 2011; Welander and Metcalf, 2008) and nuclease (*Cas9* in *Methanosarcina acetivorans*) (Nayak and Metcalf, 2017) were implemented; Genes include *Himar1/Tnp* (in *Methanococcus maripaludis,* and *Methanosarcina acetivorans*) (Zhang et al., 2000; Sattler et al., 2013), and *Tn5/Tnp* (in *Methanococcus maripaludis*) (Porat and Whitman, 2009) were used as transposons induced random mutagenesis; and *lacZ* (in *Methanococcus maripaludis*) (Gardner and Whitman, 1999), *uidA* (in *Methanosarcina* and *Methanococcus* voltae) (Pritchett et al., 2004; Beneke et al., 1995), and *bla* (in *Methanosarcina acetivorans*) (Demolli et al., 2014) were reporter genes.

engineering of biochemical pathways and metabolic engineering techniques for the enhancement of microbes for sustainable energy production.

In recent decades, the anaerobic archaea have been studied and implemented in industrial processes for wastewater and sewage treatment. However, microbiota involved in this process is more complex even within the same genus based on the substrate used. Hence, extensive research on the methanogenic community of pure and consortia of microbes are to be carried out to enhance biomethane production. Therefore, for the efficient biomethane production process, clear studies on the microorganisms and its functions are more essential to completely use organic waste. Recent advancements in the genetic implementation of methanogens are proven feasible for methanogenesis; hence, they are not economically feasible due to strict anaerobic conditions and slow growth. However, advanced genetic engineering tools are the most reasonable way to synthesize value-added products from potential methanogens.

KEYWORDS

- **bioaugmentation**
- **biomethane**
- **coenzyme B**
- **gene manipulation**
- **metabolic engineering**
- **methanogenic archaea**

REFERENCES

Argyle, J. L., Tumbula, D. L., & Leigh, J. A., (1996). Neomycin resistance as a selectable marker in *Methanococcus maripaludis*. *Appl Environ Microbiol., 62*, 4233–4237.

Asakawa, S., Akagawa-Matsushita, M., Morii, H., Koga, Y., & Hayano, K., (1995). Characterization of *Methanosarcina mazeii* TMA isolated from a paddy field soil. *Curr Microbiol., 31*, 34–38.

Awe, O. W., Zhao, Y., Nzihou, A., Minh, D. P., & Lyczko, N., (2017). A review of biogas utilization, purification and upgrading technologies. *Waste Biomass Valor., 8*, 267–283.

Azman, S., Khadem, A. F., Plugge, C. M., Stams, A. J. M., Bec, S., & Zeeman, G., (2017). Effect of humic acid on anaerobic digestion of cellulose and xylan in completely stirred tank reactors: Inhibitory effect, mitigation of the inhibition and the dynamics of the microbial communities. *Appl. Microbiol. Biotechnol., 101*, 889–901.

Balch, W. E., Fox, G. E., Magrum, L. J., Woese, C. R., & Wolfe, R. S., (1979). Methanogens: Reevaluation of a unique biological group. *Microbiol. Rev., 43*, 260–296.

Batstone, D. J., Keller, J., Angelidaki, I., Kalyuzhnyi, S. V., Pavlostathis, S. G., Rozzi, A., Sanders, W. T., et al., (2002). The IWA anaerobic digestion model no 1 (ADM1). *Water Sci. Technol., 45*, 65–73.

Beer, L., Boyd, E. S., Peters, J. W., & Posewitz, M. C., (2009). Engineering algae for biohydrogen and biofuel production, *Curr. Opin. Biotechnol., 20*, 264–271.

Belay, N., Johnson, R., Rajagopal, B. S., Conway De, M. E., & Daniels, L., (1988). Methanogenic bacteria from human dental plaque. *Appl. Environ. Microbiol., 54*, 600–603.

Beneke, S., Bestgen, H., & Klein, A., (1995). Use of the *Escherichia coli* uidA gene as a reporter in *Methanococcus voltae* for the analysis of the regulatory function of the intergenic region between the operons encoding selenium-free hydrogenases. *Mol. Gen. Gen., 248*, 225–228.

Boccazzi, P., Zhang, J. K., & Metcalf, W. W., (2000). Generation of dominant selectable markers for resistance to pseudomonic acid by cloning and mutagenesis of the *ileS* gene from the archaeon *Methanosarcina barkeri* fusaro. *J. Bacteriol., 182*, 2611–2618.

Bräuer, S. L., Cadillo-Quiroz, H., Ward, R. J., Yavitt, J. B., & Zinder, S. H., (2011). *Methanoregula boonei* gen. nov., sp. nov., an acidiphilic methanogen isolated from an acidic peat bog. *Int. J. Syst. Evol. Microbiol., 61*, 45–52.

Caruso, M. C., Braghieri, A., Capece, A., Napolitano, F., Romano, P., Galgano, F., Altieri, G., & Genovese, F., (2019). Recent updates on the use of agro-food waste for biogas production. *Appl. Sci., 9*, 1217.

Căter, M., Fanedl, L., Malovrh, S., & Marinšek, L. R., (2015). Biogas production from brewery spent grain enhanced by bioaugmentation with hydrolytic anaerobic bacteria. *Bioresour. Technol., 186*, 261–269.

Cirne, D. G., Paloumet, X., Bjornsson, L., Alves, M. M., & Mattiasson, B., (2007). Anaerobic digestion of lipid-rich waste—effects of lipid concentration. *Renew. Energy, 32*, 965–975.

Daniels, L., Fuchs, G., Thauer, R. K., & Zeikus, J. G., (1977). Carbon monoxide oxidation by methanogenic bacteria. *J. Bacteriol., 132*, 118–126.

Demolli, S., Geist, M. M., Weigand, J. E., Matschiavelli, N., Suess, B., & Rother, M., (2014). Development of β-lactamase as a tool for monitoring conditional gene expression by a tetracycline-riboswitch in *Methanosarcina acetivorans*. *Archaea*, 1–10.

Doudna, J. A., & Charpentier, E., (2014). The new frontier of genome engineering with CRISPR-Cas9. *Science, 346*, 1258096.

Dridi, B., Fardeau, M. L., Ollivier, B., Raoult, D., & Drancourt, M., (2012). *Methanomassiliicoccus luminyensis* gen. nov., sp. nov., a methanogenic archaeon isolated from human faeces. *Int. J. Syst. Evol. Microbiol., 62*, 1902–1907.

Enzmann, F., Mayer, F., Rother, M., & Holtmann, D., (2018). Methanogens: Biochemical background and biotechnological applications. *AMB Express, 8*(1).

Evans, P. N., Parks, D. H., Chadwick, G. L., Robbins, S. J., Orphan, V. J., Golding, S. D., & Tyson, G. W., (2015). Methane metabolism in the archaeal phylum *Bathyarchaeota* revealed by genome-centric metagenomics. *Science, 350*, 434–438.

Ferrari, A., Brusa, T., Rutili, A., Canzi, E., & Biavati, B., (1994). Isolation and characterization of *Methanobrevibacter oralis* sp. nov. *Curr. Microbiol., 29*, 7–12.

Fotidis, I. A., Wang, H., Fiedel, N. R., Luo, G., & Karakashev, D. B., (2014). Bioaugmentation as a solution to increase methane production from an ammonia-rich substrate. *Environ. Sci. Technol., 48*, 7669–7676.

Garcia, J. L., Patel, B. K., & Ollivier, B., (2000). Taxonomic, phylogenetic, and ecological diversity of methanogenic *Archaea*. *Anaerobe., 6*, 205–226.

Gardner, W. L., & Whitman, W. B., (1999). Expression vectors for *Methanococcus maripaludis*: Overexpression of acetohydroxyacid synthase and beta-galactosidase. *Genetics, 152*, 1439–1447.

Gernhardt, P., Possot, O., Foglino, M., Sibold, L., & Klein, A., (1990). Construction of an integration vector for use in the archaebacterium *Methanococcus voltae* and expression of a eubacterial resistance gene. *Mol. Gen. Genet., 221*, 273–279.

Guss, A. M., Mukhopadhyay, B., Zhang, J. K., & Metcalf, W. W., (2005). Genetic analysis of much mutants in two *Methanosarcina* species demonstrates multiple roles for the methanopterin-dependent C-1 oxidation/reduction pathway and differences in H_2 metabolism between closely related species. *Mol. Microbiol., 55*, 1671–1680.

Guss, A. M., Rother, M., Zhang, J. K., Kulkarni, G., & Metcalf, W. W., (2008). New methods for tightly regulated gene expression and highly efficient chromosomal integration of cloned genes for *Methanosarcina* species. *Archaea, 2*, 193–203.

Harris, J. E., & Pinn, P. A., (1985). Bacitracin-resistant mutants of a mesophilic *Methanobacterium* species. *Arch. Microbiol., 143*, 151–153.

Ho, S. H., Huang, S. W., Chen, C. Y., Hasunuma, T., Kondo, A., & Chang, J. S., (2013). Bioethanol production using carbohydrate-rich microalgae biomass as feedstock. *Bioresour. Technol., 135*, 191–198.

Hohn, M. J., Palioura, S., Su, D., Yuan, J., & Söll, D., (2011). Genetic analysis of selenocysteine biosynthesis in the archaeon *Methanococcus maripaludis*. *Mol. Microbiol., 81*, 249–258.

Holm-Nielsen, J. B., Al Seadi, T., & Oleskowicz-Popiel, P., (2009). The future of anaerobic digestion and biogas utilization. *Bioresour. Technol., 100*, 5478–5484.

Hummel, H., & Böck, A., (1985). Mutations in *Methanobacterium formicicum* conferring resistance to anti-80S ribosome-targeted antibiotics. *Mol. Gen. Genet., 198*, 529–533.

Jaenicke, S., Zakrzewski, M., Junemann, S., Puhler, A., Groesmann, A., & Schluter, A., (2011). Analysis of the metagenome from biogas-producing microbial community by means of bioinformatics methods. In: De Bruijn, F. J., (ed.), *Handbook of Molecular Microbial Ecology, Vol. II: Metagenomics in Different Habitats*. Wiley-Blackwell, Hoboken.

Jin, H., Liu, R., & He, Y., (2012). Kinetics of batch fermentations for ethanol production with immobilized *Saccharomyces cerevisiae* growing on sweet sorghum stalk juice. *Proc. Environ. Sci., 12*, 137–145.

Jones, W. J., Leigh, J. A., Mayer, F., Woese, C. R., & Wolfe, R. S., (1983). *Methanococcus jannaschii* sp. nov., an extremely thermophilic methanogen from a submarine hydrothermal vent. *Arch. Microbiol., 136*, 254–261.

Karakashev, D., Batstone, D. J., Trably, E., & Angelidaki, I., (2006). Acetate oxidation is the dominant methanogenic pathway from acetate in the absence of *Methanosaetaceae*. *Appl. Environ. Microbiol., 72*, 5138–5141.

Kárászová, M., Sedláková, Z., & Izák, P., (2015). Gas permeation processes in biogas upgrading: A short review. *Chem. Pap., 69*, 1277–1283.

Kern, T., Fischer, M. A., Deppenmeier, U., Schmitz, R. A., & Rother, M., (2016). *Methanosarcina flavescens* sp. nov., a methanogenic archaeon isolated from a full-scale anaerobic digester. *Int. J. Syst. Evol. Microbiol., 66*, 1533–1538.

Kern, T., Linge, M., & Rother, M., (2015). *Methanobacterium aggregans* sp. nov., a hydrogenotrophic methanogenic archaeon isolated from an anaerobic digester. *Int. J. Syst. Evol. Microbiol., 65*, 1975–1980.

Klein, A., & Horner, K., (1995). Integration vectors for *Methanococci*. In: Sowers, K. R., & Schreier, H. J., (eds.), *Methanogens* (pp. 409–411). Cold Spring Harbor Laboratory Press.

Kougias, P. G., Treu, L., Benavente, D. P., Boe, K., Campanaro, S., & Angelidaki, I., (2017). Ex-situ biogas upgrading and enhancement in different reactor systems. *Bioresour. Technol., 225*, 429–437.

Kurr, M., Huber, R., König, H., Jannasch, H. W., Fricke, H., Trincone, A., Kristjansson, J. K., & Stetter, K. O., (1991). *Methanopyrus kandleri*, gen. and sp. nov. represents a novel group of hyperthermophilic methanogens, growing at 110°C. *Arch. Microbiol., 156*, 239–247.

Lee, S. J., Lee, S. J., & Lee, D. W., (2013). Design and development of synthetic microbial platform cells for bioenergy. *Front. Microbiol., 4*, 92.

Lie, T. J., & Leigh, J. A., (2003). A novel repressor of *nif* and *glnA* expression in the methanogenic archaeon *Methanococcus maripaludis*. *Mol. Microbiol., 47*, 235–246.

Liu, Y., & Whitman, W. B., (2008). Metabolic, phylogenetic, and ecological diversity of the methanogenic archaea. *Ann. NY Acad. Sci., 1125*, 171–189.

Mathrani, I. M., Boone, D. R., Mah, R. A., Fox, G. E., & Lau, P. P., (1988). *Methanohalophilus zhilinae* sp. nov., an alkaliphilic, halophilic, methylotrophic methanogen. *Int. J. Syst. Bacteriol., 38*, 139–142.

McInerney, M. J., Struchtemeyer, C. G., Sieber, J., Mouttaki, H., Stams, A. J., Schink, B., Rohlin, L., & Gunsalus, R. P., (2008). Physiology, ecology, phylogeny, and genomics of microorganisms capable of syntrophic metabolism. *Ann. NY Acad. Sci., 1125*, 58–72.

Metcalf, W. W., Zhang, J. K., Apolinario, E., Sowers, K. R., & Wolfe, R. S., (1997). A genetic system for Archaea of the genus *Methanosarcina*: Liposome-mediated transformation and construction of shuttle vectors. *PNAS, 94*, 2626–2631.

Metcalf, W. W., Zhang, J. K., Shi, X., & Wolfe, R. S., (1996). Molecular, genetic, and biochemical characterization of the *serC* gene of *Methanosarcina barkeri* fusaro. *J. Bacteriol., 178*, 5797–5802.

Miller, T. L., Wolin, M. J., Conway De, M. E., & Macario, A. J., (1982). Isolation of *Methanobrevibacter smithii* from human feces. *Appl. Environ. Microbiol., 43*, 227–232.

Mochimaru, H., Tamaki, H., Hanada, S., Imachi, H., Nakamura, K., Sakata, S., & Kamagata, Y., (2009). *Methanolobus profundi* sp. nov., a methylotrophic methanogen isolated from deep subsurface sediments in a natural gas field. *Int. J. Syst. Evol. Microbiol., 59*, 714–718.

Mondorf, S., Deppenmeier, U., & Welte, C., (2012). A novel inducible protein production system and neomycin resistance as selection marker for *Methanosarcina mazei*. *Archaea*, 1–8.

Moore, B. C., & Leigh, J. A., (2005). Markerless mutagenesis in *Methanococcus maripaludis* demonstrates roles for alanine dehydrogenase, alanine racemase, and alanine permease. *J. Bacteriol., 187*, 972–979.

Mukhopadhyay, A., (2015). Tolerance engineering in bacteria for the production of advanced biofuels and chemicals. *Trends Microbiol., 23*, 498–508.

Nanda, S., & Berruti, F., (2021a). A technical review of bioenergy and resource recovery from municipal solid waste. *J. Hazard. Mater., 403*, 123970.

Nanda, S., & Berruti, F., (2021b). Municipal solid waste management and landfilling technologies: A review. *Environ. Chem. Lett*, 19, 1433–1456.

Nanda, S., & Berruti, F., (2021c). Thermochemical conversion of plastic waste to fuels: A review. *Environ. Chem. Lett.*, 19, 123–148.

Nanda, S., Azargohar, R., Dalai, A. K., & Kozinski, J. A., (2015). An assessment on the sustainability of lignocellulosic biomass for biorefining. *Renew. Sust. Energy Rev., 50*, 925–941.

Nanda, S., Golemi-Kotra, D., McDermott, J. C., Dalai, A. K., Gökalp, I., & Kozinski, J. A., (2017). Fermentative production of butanol: Perspectives on synthetic biology. *New Biotechnol., 37*, 210–221.

Nanda, S., Mohammad, J., Reddy, S. N., Kozinski, J. A., & Dalai, A. K., (2014). Pathways of lignocellulosic biomass conversion to renewable fuels. *Biomass Conv. Bioref., 4*, 157–191.

Nayak, D. D., & Metcalf, W. W., (2017). Cas9-mediated genome editing in the methanogenic archaeon *Methanosarcina acetivorans*. *PNAS, 114*, 2976–2981.

Nzila, A., (2017). Mini-review: Update on bioaugmentation in anaerobic processes for biogas production. *Anaerobe, 46*, 3–17.

O'Brien, J. M., Wolkin, R. H., Moench, T. T., Morgan, J. B., & Zeikus, J. G., (1984). Association of hydrogen metabolism with unitrophic or mixotrophic growth of *Methanosarcina barkeri* on carbon monoxide. *J. Bacteriol., 158*, 373–375.

Oelgeschläger, E., & Rother, M., (2009). In vivo role of three fused corrinoid/ methyl transfer proteins in *Methanosarcina acetivorans*. *Mol. Microbiol., 72*, 1260–1272.

Okolie, J. A., Nanda, S., Dalai, A. K., & Kozinski, J. A., (2021). Chemistry and specialty industrial applications of lignocellulosic biomass. *Waste Biomass Valor.*, 12, 2145–2169.

Okolie, J. A., Nanda, S., Dalai, A. K., Berruti, F., & Kozinski, J. A., (2020). A review on subcritical and supercritical water gasification of biogenic, polymeric and petroleum wastes to hydrogen-rich synthesis gas. *Renew. Sust. Energy Rev., 119*, 109546.

Öner, B. E., Akyol, C., Bozan, M., Ince, O., Aydin, S., & Ince, B., (2018). Bioaugmentation with *Clostridium thermocellum* to enhance the anaerobic biodegradation of lignocellulosic agricultural residues. *Bioresour. Technol., 249*, 620–625.

Papilo, P., Kusumanto, I., & Kunaifi, K., (2017). Assessment of agricultural biomass potential to electricity generation in Riau province. *IOP Conf. Series: Earth Environ. Sci., 65*, 012006.

Patterson, T., Esteves, S., Dinsdale, R., & Guwy, A., (2011). An evaluation of the policy and techno-economic factors affecting the potential for biogas upgrading for transport fuel use in the UK. *Energy Policy, 39*, 1806–1816.

Peng, X., Börner, R. A., Nges, I. A., & Liu, J., (2014). Impact of bioaugmentation on biochemical methane potential for wheat straw with addition of *Clostridium cellulolyticum*. *Bioresour. Technol., 152*, 567–571.

Porat, I., & Whitman, W. B., (2009). Tryptophan auxotrophs were obtained by random transposon insertions in the *Methanococcus maripaludis* tryptophan operon. *FEMS Microbiol. Lett., 297*, 250–254.

Poszytek, K., Ciezkowska, M., Sklodowska, A., & Drewniak, L., (2016). Microbial Consortium with High Cellulolytic Activity (MCHCA) for enhanced biogas production. *Front. Microbiol., 7*, 324.

Pritchett, M. A., & Metcalf, W. W., (2005). Genetic, physiological and biochemical characterization of multiple methanol methyltransferase isozymes in *Methanosarcina acetivorans* C2A. *Mol. Microbiol., 56*, 1183–1194.

Pritchett, M. A., Zhang, J. K., & Metcalf, W. W., (2004). Development of a markerless genetic exchange method for *Methanosarcina acetivorans* C2A and its use in construction of new genetic tools for methanogenic archaea. *Appl. Environ. Microbiol., 70*, 1425–1433.

Raboni, M., & Urbini, G., (2014). Production and use of biogas in Europe: A survey of current status and perspectives. *Ambient. Agua., 9*.

Ren, T., Patel, M., & Blok, K., (2008). Steam cracking and methane to olefins: Energy use, CO_2 emissions and production costs. *Energy, 33*, 817–833.

Risberg, K., Cederlund, H., Pell, M., Arthurson, V., & Schnturer, A., (2017). Comparative characterization of digestate versus pig slurry and cow manure—chemical composition and effects on soil microbial activity. *Waste Manag., 61*, 529–538.

Robichaux, M., Howell, M., & Boopathy, R., (2003). Methanogenic activity in human periodontal pocket. *Curr. Microbiol., 46*, 53–58.

Rother, M., & Metcalf, W. W., (2005). Genetic technologies for archaea. *Curr. Opin. Microbiol., 8*, 745–751.

Sarmiento, F. B., Leigh, J. A., & Whitman, W. B., (2011). *Genetic Systems for Hydrogeno-Trophic Methanogens*, 43–73.

Sasaki, K., Morita, M., Sasaki, D., Nagaoka, J., Matsumoto, N., Pohmura, N., et al., (2011). Syntrophic degradation of proteinaceous materials by the thermophilic strains *Coprothermobacter proteolyticus* and *Methanothermobacter thermautotrophicus*. *J. Biosci. Bioeng., 112*, 469–472.

Sattler, C., Wolf, S., Fersch, J., Goetz, S., & Rother, M., (2013). Random mutagenesis identifies factors involved in formate-dependent growth of the methanogenic archaeon *Methanococcus maripaludis*. *Mol. Genet. Genom., 288*, 413–424.

Schink, B., Ward, J. C., & Zeikus, J. G., (1981). Microbiology of wet wood: Importance of pectin degradation and *Clostridium* species in living trees. *Appl. Environ. Microbiol., 42*, 526–532.

Schnürer, A., (2016). Biogas production: Microbiology and technology. In: Hatti-Kaul, R., Mamo, G., & Mattiasson, B., (eds.), *Advances in Biochemical Engineering/Biotechnology* (Vol. 156, pp. 195–234). Switzerland: Springer International Publishing.

Senthilkumar, V., & Gunasekaran, P., (2005). Bioethanol production from cellulosic substrates: Engineered bacteria and process integration challenges. *J. Sci. Ind. Res., 64*, 845–853.

Silva, S. A., Cavaleiro, A. J., Pereira, M. A., Stams, A. J. M., Alves, M. M., & Sousa, D. Z., (2014). Long-term acclimation of anaerobic sludges for high-rate methanogenesis from LCFA. *Biomass Bioenergy, 67*, 297–303.

Solli, L., Håvelsrud, O. E., Horn, S. J., & Rike, A. G., (2014). A metagenomic study of the microbial communities in four parallel biogas reactors. *Biotechnol. Biofuels, 7*, 146.

Speda, J., Johansson, M. A., Odnell, A., & Karlsson, M., (2017). Enhanced biomethane production rate and yield from lignocellulosic ensiled forage ley by in situ anaerobic digestion treatment with endogenous cellulolytic enzymes. *Biotechnol. Biofuels, 10*, 129.

Srivastava, R. K., (2019). Bioenergy production by contribution of effective and suitable microsystem. *Mater. Sci. Sci. Technol., 2*, 308–318.

Strong, P. J., Xie, S., & Clarke, W. P., (2015). Methane as a resource: Can the methanotrophs add value? *Environ. Sci. Technol., 49*, 4001–4018.

Sun, J., & Klein, A., (2004). A lysR-type regulator is involved in the negative regulation of genes encoding selenium-free hydrogenases in the archaeon *Methanococcus voltae*. *Mol. Microbiol., 52*, 563–571.

Sun, L., Liu, T., Muller, B., & Schnurer, A., (2016). The microbial community structure in industrial biogas plants influences the degradation rate of straw and cellulose in batch tests. *Biotechnol. Biofuels, 9*, 128.

Thauer, R. K., Kaster, A. K., Seedorf, H., Buckel, W., & Hedderich, R., (2008). Methanogenic archaea: Ecologically relevant differences in energy conservation. *Nat. Rev. Microbiol., 6*, 579–591.

Tumbula, D. L., Bowen, T. L., & Whitman, W. B., (1997). Characterization of pURB500 from the archaeon *Methanococcus maripaludis* and construction of a shuttle vector. *J. Bacteriol., 179*, 2976–2986.

Vanwonterghem, I., Evans, P. N., Parks, D. H., Jensen, P. D., Woodcroft, B. J., Hugenholtz, P., & Tyson, G. W., (2016). Methylotrophic methanogenesis discovered in the archaeal phylum *Verstraetearchaeota*. *Nat. Microbiol., 1*, 16170.

Vavilin, V. A., Fernandez, B., Palatsi, J., & Flotats, X., (2008). Hydrolysis kinetics in anaerobic degradation of particulate organic material: An overview. *Waste Manag., 28*, 939–951.

Wagner, T., Ermler, U., & Shima, S., (2016). The methanogenic CO_2 reducing-and- fixing enzyme is bifunctional and contains 46 [4Fe–4S] clusters. *Science, 354*, 114–117.

Wang, M., Zhou, J., Yuan, Y. X., Dai, Y. M., Li, D., Li, Z. D., et al., (2017). Methane production characteristics and microbial community dynamics of mono-digestion and codigestion using corn stalk and pig manure. *Int. J. Hydrogen Energy, 42*, 4893–4901.

Wang, S., Hou, X., & Su, H., (2017). Exploration of the relationship between biogas production and microbial community under high salinity conditions. *Sci. Rep., 7*, 1149.

Weiland, P., (2010). Biogas production: Current state and perspectives. *Appl. Microbiol. Biotechnol., 85*, 849–860.

Welander, P. V., & Metcalf, W. W., (2005). Loss of the mtr operon in *Methanosarcina* blocks growth on methanol, but not methanogenesis, and reveals an unknown methanogenic pathway. *PNAS, 102*, 10664–10669.

Welander, P. V., & Metcalf, W. W., (2008). Mutagenesis of the C1 oxidation pathway in *Methanosarcina barkeri*: New insights into the Mtr/Mer bypass pathway. *J. Bacteriol., 190*, 1928–1936.

Wirth, R., Kovacs, E., Maroti, G., Bagi, Z., Rakhely, G., & Kovacs, K. L., (2012). Characterization of a biogas-producing microbial community by short-read next generation DNA sequencing. *Biotechnol. Biofuels, 5*, 41.

Worrell, V. E., Nagle, D. P., McCarthy, D., & Eisenbraun, A., (1988). Genetic transformation system in the archaebacterium *Methanobacterium thermoautotrophicum* Marburg. *J. Bacteriol., 170*, 653–656.

Xia, A., Cheng, J., & Murphy, J. D., (2016). Innovation in biological production and upgrading of methane and hydrogen for use as gaseous transport biofuel. *Biotechnol. Adv., 34*, 451–472.

Zeikus, J. G., & Henning, D. L., (1975). *Methanobacterium arbophilicum* sp. nov. An obligate anaerobe isolated from wetwood of living trees. *Antonie Van Leeuwenhoek, 41*, 543–552.

Zhang, J. K., Pritchett, M. A., Lampe, D. J., Robertson, H. M., & Metcalf, W. W., (2000). *In vivo* transposon mutagenesis of the methanogenic archaeon *Methanosarcina acetivorans* C2A using a modified version of the insect mariner-family transposable element Himar1. *PNAS, 97*, 9665–9670.

Zhang, J., Guo, R., Qiu, Y., Qiao, J., Yuan, X., Shi, X., & Wang, C.S., (2015). Bioaugmentation with an acetate-type fermentation bacterium *Acetobacteroides hydrogenigenes* improves methane production from corn straw. *Bioresour. Technol., 179*, 306–313.

CHAPTER 4

Biomethane Production through Anaerobic Digestion of Lignocellulosic Biomass and Organic Wastes

ALIVIA MUKHERJEE,[1] BISWA R. PATRA,[1] FALGUNI PATTNAIK,[2] JUDE A. OKOLIE,[1] SONIL NANDA,[1] and AJAY K. DALAI[1]

[1]*Department of Chemical and Biological Engineering, University of Saskatchewan, Saskatoon, Saskatchewan, Canada*
E-mail: ajay.dalai@usask.ca (Ajay K. Dalai)

[2]*Center for Rural Development and Technology, Indian Institute of Technology Delhi, New Delhi, India*

ABSTRACT

Biogas generation via anaerobic digestion of lignocellulosic biomass entails a developed technology and a cost-effective pathway. The technology may add further appeal to lignocellulosic waste management owing to the surplus availability and nonedibility of the precursors in rural areas. The biogas generated from anaerobic digestion contributes significantly to the ever-increasing demand for sustainable and affordable energy and products. The current review is an attempt to present the state-of-the-art of anaerobic digestion of biogenic solid wastes, the impact of different chemical pretreatment processes and products and the influence of operating parameters on biomethane yields. In this regard, various pretreatment strategies like acidic, alkaline, oxidative, and ionic liquid have been developed to overcome the inherent recalcitrance and complex crystallinity of cellulose and lignin to simpler forms like monomer. The present review highlights the use of chemical agents including mild alkali, dilute acid, ionic liquid, and oxidants which trigger solubilization and degradation of lignocellulosic wastes and utilization for enhanced biogas generation than the pristine substrate. The comparison of different chemical pretreatment strategies for enhanced biogas yield is presented. Additionally, the influence of the operational parameters like temperature, pH, reactor type, hydraulic retention time, carbon/nitrogen ratio and chemical loading on the performance of anaerobic digestion towards the quality and quantity of biogas yield are inherently reviewed. Lastly, the emphasis is also placed on the global biomethane market towards a sustainable future.

4.1 INTRODUCTION

Sustainable energy generation from alternate and renewable energy sources is a subject widely discussed both in public forums and academic spaces because of the unremitting development of countries. The need for the diversification and expansion of sustainable energy is increasing because of the depletion of fossil fuel, the growing unsustainability of the old-linear economy coupled with the threat of climate change due to greenhouse gas (GHG) emissions. In this regard, anaerobic digestion (AD) has become a promising approach to produce biomethane from virtually all diverse biogenic solid wastes. It is gaining attention due to several noteworthy merits including the use of abundantly available bioresources, clean energy production, reduced greenhouse gas emission, applicability in remote areas, and widespread domestic applications of biomethane over conventional nonrenewable sources (Moghaddam et al., 2019; Oechsner et al., 2015; Angelidaki et al., 2011). Anaerobic digestion, one class of a biochemical process executed in the absence of oxygen in a closed digester or sealed lagoon, in which lignocellulosic rich organic substrates are mediated by a microbial consortium to produce biogas and digestate (Mittal et al., 2018; El-Mashad, 2013). The exploitation of biogas and the digestate further as a renewable source of energy and biofertilizer has gained attention in recent years (Shetty et al., 2020; Pellera and Gidarakos, 2018).

In recent years to produce an alternative renewable source of energy such as biomethane, lignocellulosic biomass, a ubiquitous source of energy has been under the radar as an appropriate substrate (Yadav et al., 2019; Aziz et al., 2019; Tapadia-Maheshwari et al., 2019; Neshat et al., 2017). For renewable energy generation, lignocellulosic biomasses such as agricultural and forest residues, energy crops and a part of municipal solid waste could be explored because of their abundance, carbon-neutrality, and low cost (Gu et al., 2015; Saini et al., 2015). The acceptance rate of using lignocellulosic biomass like agriculture residues as feedstock for biomethane generation is considerably high owing to its maximum biogas yield with less energy input (Okolie et al., 2020, 2021).

The structural composition of lignocellulose based biomass is different from other organic materials because the earlier one posses a strong and tight structure attributed to the presence of cellulose (40–60 wt.%), followed by hemicellulose (10–40 wt.%) and lignin (10–30 wt.%) in a cell wall (Nanda et al., 2015; Yadav et al., 2019). The main components of lignocellulosic biomass, i.e., cellulose, and hemicellulose are composed of potentially fermentable sugars which via anaerobic digestion technique could be valorized into biogas. Although, among all the components, lignin is the most recalcitrant compound, hindering the overall efficacy of biomethane generation with low biodegradability. The biomethane productivity from pristine lignocellulosic biomass is not large owing to its recalcitrant structural arrangement and composition that limits the hydrolysis step and the formation of hydrolyzable sugars. However, a pretreatment phase could amplify the generation of biomethane and improve the overall production efficiency before subjecting the lignocellulosic biomass to anaerobic digestion (Yadav et al., 2019; Dollhofer et al., 2018). With pretreatment, the structure of the cell wall is destroyed, which then converts the cellulose and hemicellulose into digestible fractions and remove some traces of lignin which further amplifies the generation of biomethane (Lindner et al., 2015; Hirunsupachote and Chavalparit, 2019).

Although, losses of valuable organic material or formation of inhibitors could take place with intense pretreatment of the substrate (Lindner et al., 2015; Sivagurunathan et al., 2017). Various pretreatment techniques can be carried out by implementing some cost-effective and simple methods like chemical (mild alkali, dilute acid or ionic liquids) and biological pretreatment (fungal) (Bhutto et al., 2017; Rouches et al., 2019) and some involves considerable energy like thermal (liquid hot water or steam explosion) or mechanical (milling or grinding) pretreatment techniques (Kim et al., 2018). Numerous lab-scale chemical pretreatment of the substrate over biological, thermochemical or mechanical have been carried out and found to be the most investigated technique in the literature. The application of environmentally friendly and economically viable pretreatment techniques of lignocellulosic biomass for biomethane generation is the subject of much scientific community.

Additionally, many operating parameters like the temperature of the digester or sealed lagoon, pH, C/N ratio, chemical loading or type of the reactor could influence the overall generation of biogas. Optimizing the anaerobic digestion method would highly rely on how carefully the operating parameters are chosen, and a small variation could make a big difference in industrial processing facilities (Sarker et al., 2019; Ngumah et al., 2017).

Biomethanation process has gained great attraction in the scientific community for effectively addressing the environmental challenges like waste disposal and valorization of various organic residues into value-added products such as vehicle fuel, biogas, biomethane, and macro and micronutrients-rich organic fertilizer suitable as organic fertilizer (Sarker et al., 2019; Tapadia-Maheshwari et al., 2019; Mishra et al., 2018; Zhang et al., 2015). An energy-rich carrier such as biogas, a potential candidate for diverse downstream conversion technology, is considered as one of the foundations of renewable energy sources, which has already found applications in several countries (Koupaie et al., 2019).

Biogas, an amalgamation of mainly CH_4 (45–75%), CO_2 (25–55%) and a small amount of impurities (< 3%), can be exploited to produce electricity as it requires to contain more than 40% of methane to be flammable (Ngumah et al., 2017). Moreover, it possesses advantages over conventional non-renewable resources, which presently has a declining trend. It has paved the way to its widespread enormous applications such as (i) cooking fuel in developing countries, (ii) could be used for electricity generation in large scale plants, (iii) further upgraded to potential automobile-fuel or compressed natural gas purification (bio-CNG) and (iv) injected in the natural gas network after (Tapadia-Maheshwari et al., 2019; Hirunsupachote and Chavalparit 2019; Koupaie et al., 2019; Tayyab et al., 2018). Biogas has diverse energetic applications. The application of products and byproducts from various stages of biomethanation is illustrated in Figure 4.1. In this diagram, the overview of the biogenic solid wastes as the potential feedstock for anaerobic digestion, chemical pretreatment technique before anaerobic digestion, and the influence of operational parameters and their interdependence to the overall efficiency on biomethane generation are discussed. This chapter aims to discuss and emphasize the ways to bring innovations for a future clean source of energy.

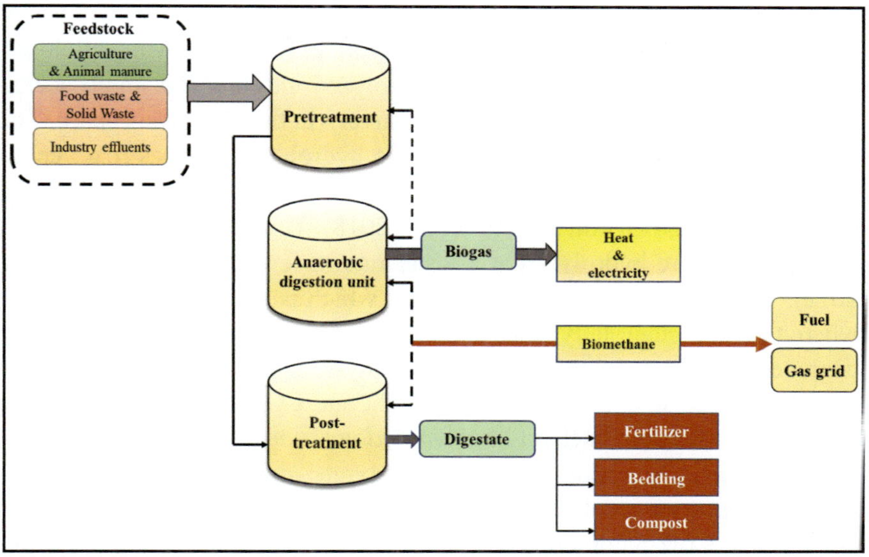

FIGURE 4.1 Energy products from the conversion of feedstock through biomethanation.

4.2 BIOLOGICAL METHANE PRODUCTION

Biomethane, or popularly known as biogas, is produced through both thermocatalytic and biological processes. However, the biological production of methane from several wastes and lignocellulosic biomasses is becoming popularized due to its potentiality towards the economical processing costs. Biological production of methane comprises several stages, which majorly depend upon the types of feedstock implemented for the methane production. Generally, this process of methane production is known as methanogenesis, which is mainly carried out in an anaerobic atmosphere by methanogens (methane-producing microorganisms).

The biological methane production is subcategorized into four steps of anaerobic digestion, such as hydrolysis of the complex organic molecules, degradation of long-chain fatty acids, acetogenesis of organic acids and alcohols, and formation of methane (methanogenesis). These phases of the methane production process are carried out synchronously by the fermentative and symbiotic action of mixed anaerobic microorganisms (Pattanaik et al., 2019). Each of the phases of the methane production has its contribution towards the whole process, and all phases proceed in a chain mechanism, i.e., phases are interrelated.

Usually, the feedstocks used for the methane production are in the form of complex organic molecules such as carbohydrates (cellulose, hemicelluloses, etc.), proteins, and lipids which get hydrolyzed into mono or oligosaccharides, amino acids, and fatty acids, respectively, at the hydrolysis phase. The process of hydrolysis is performed by the hydrolytic enzymes (e.g., cellulases, lipases, and proteases) secreted from the various anaerobic bacteria (Molino et al., 2013; Raveendran et al., 2018). Unlike proteins and lipids, carbohydrates are hydrolyzed at a faster rate in few hours (Chandra et al., 2012). In the second stage of the process which is named the acidogenesis where the long-chain fatty acids and other produced mono or oligomers are degraded into short-chain fatty acids like butyric acid, propionic acid, formic acid, and acetic acid; alcohols, acetates, H_2, and CO_2 in the influence of the acidogenic anaerobic bacteria like *Clostridium, Lactobacillus, Escherichia, Streptococcus, Sytrophomonoswolfei* (butyrate decomposer), *Syntrophomonoswolinii* (propionate decomposer), etc., (Anderson et al., 2003; Molino et al., 2013; Cruz-Salomón et al., 2020). These fragmented molecules after acidogenesis play the substrate for the acetogenesis stage where H_2, CO_2, acetic acid, and alcohols are converted to acetate, formate, butyrate or propionate (ester form of the organic acids) form in the presence of acetogenic bacteria (collectively known as acetogens). The esters (majorly acetates or formate) and gases like CO_2 and H_2 are fermented to form a mixture of gases and water by the methanogens (*Methanogenium, Methanobacillus, Methanosarcina, Methanococcus*, etc.), where CO_2 constitutes 25–45% and CH_4 constitutes 55–75% (Pattanaik et al., 2019; Chandra et al., 2012).

In methanogenesis, *Methanosarcina*, and *Methanococcus* are considered as highly efficient methanogens, which can digest both CO_2 or H_2; and acetates anaerobically to produce methane (Molino et al., 2013; Holmes and Smith, 2016). The whole process comprising of different phases and the reactions involved in the biomethanation process are diagrammatically illustrated in Figure 4.2 and Eq. (4.1–4.13) (Molino et al., 2013; Schmidt and Ahring, 1993).

$$\text{Carbohydrates} \rightarrow \text{Mono or oligosaccharides} \tag{4.1}$$

$$\text{Protein} \rightarrow \text{Amino acids} \tag{4.2}$$

$$\text{Lipid} \rightarrow \text{Fatty acids} \tag{4.3}$$

$$\text{Long chain fatty acids} \rightarrow \text{Butyric, Propionic, Acetic, and Formic acid} \tag{4.4}$$

$$C_6H_{12}O_6 \text{ (Glucose)} \rightarrow 2CH_3CH_2OH + 2CO_2 \tag{4.5}$$

$$\text{Organic acids} \rightarrow \text{Butyrate, Propionate, Acetate, and Formate} \quad (4.6)$$

$$CH_3\ CH_2\ CH_2\ COO^- \text{(Butyrate)} + 2H_2O \rightarrow 2CH_3\ COO^- \text{(Acetate)} + 2H_2 + H^+ \quad (4.7)$$

$$CH_3\ CH_2\ CH_2\ COO^- + 2HCO_3^- \rightarrow 2CH_3\ COO^- + 2HCOO^- \text{(Formate)} + H^+ \quad (4.8)$$

$$CH_3\ CH_2\ COO^- \text{(Propionate)} + 3H_2O \rightarrow CH_3\ COO^- + HCO_3^- + 3H_2 + H^+ \quad (4.9)$$

$$CH_3\ CH_2\ COO^- + 2HCO_3^- \rightarrow CH_3\ COO^- + 3HCOO^- + H^+ \quad (4.10)$$

$$CH_3\ COO^- + H_2O \rightarrow CH_4 + HCO_3^- \quad (4.11)$$

$$4H_2 + HCO_3^- + H^+ \rightarrow CH_4 + 3H_2O \quad (4.12)$$

$$4HCOO^- + H_2\ O^- + H^+ \rightarrow CH_4 + 3HCO_3^- \quad (4.13)$$

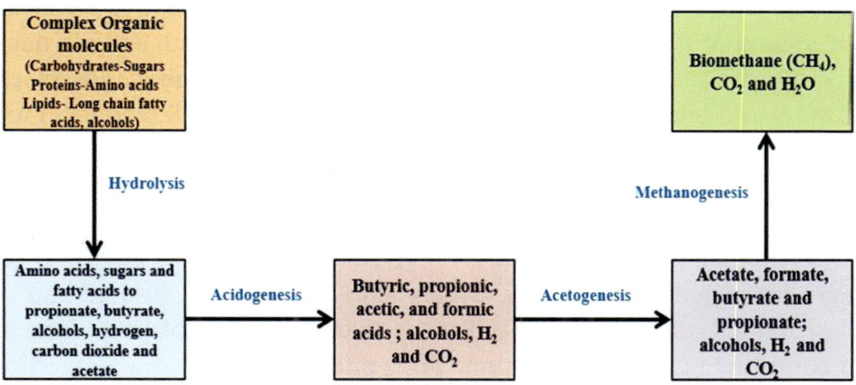

FIGURE 4.2 Different stages in anaerobic digestion.

4.3 BIOGENIC SOLID WASTE

There was a time when biogas or biomethane was used to be produced exclusively from the cow dung in many parts of India. Biomethanation is carried out using feedstocks with moisture content typically greater than 50% (Yu and Schanbacher, 2010). The feedstocks for biomethane production can be mainly categorized by municipal solid waste, agro-based wastes, energy crops, and animal wastes. Solid wastes collectively contain household wastes, biomedical wastes, sewage waste, and other various wastes generated from different sectors in a municipality (Nanda and Berruti, 2021). Municipal solid waste contains a generous amount of water contents and various anaerobic methanogenic bacteria, which makes it a potential substrate to produce biomethane. However, methane production from the municipal solid waste varies from 85–390 mL per dry ton of the substrate, which is comparatively of lower yield as compared to the other wastes (Yu et al., 2010).

Negi et al. (2018) took the mixed feedstocks containing municipal solid waste and rice straw where the co-digestion process increased the biomethane potential by 57%. Another type of potential feedstock, animal wastes comprise of animal dung and fodder wastes like residual straw, hay, waste leafy items. Biogas or biomethane from the cow dung is already popularized in villages in India due to its larger availability over there. Furthermore, there are several instances of co-digestion of animal and poultry manure with water hyacinth and agro-based residues like maize silage, where the biochemical methane potential (BMP) was found to be around 55% for the co-digestion of freshwater hyacinth and pig dung (Sukasem and Prayoonkham, 2017). Moreover, in the above instances, there was a hike in the methane production yield from 5800 to 6580 m^3/day in the co-digestion of maize silage with the cattle and poultry manure, which is equivalent to the energy value of 34,600 kWh (Yangin-Gomec and Ozturk, 2013).

Agricultural wastes consist of crop residue and waste from the agro-based industries. A few studies have reported the co-digestion of agriculture waste with other substrates, such as the use of fruit waste as a cosubstrate of the chicken rumen (Ogunleye et al., 2016), co-digestion of various vegetable wastes with other kinds of organic wastes (Patil and Deshmukh, 2015), co-digestion of vegetable and fruit waste (Prakash and Singh, 2013).

Many other instances are available in the literature. Besides the co-digestion process, researchers have documented several works on the use of a single substrate to produce biomethane. Shetty et al. (2017) took rice straw as the substrate to produce biogas or biomethane with alkali pretreatment. They found the biogas yield of 514 L/kg with a methane content of 59%.

Okeh et al. (2014) optimized the biomethanation process taking rice husk as the feedstock and concluded with a biogas yield of 383 mL/day with pH of the digestion medium equals to 7, water to feed ratio of 6:1 and by accompanying with the heavy metals (zinc, nickel, and copper) in the digester to catalyze the process. In this way, agro-based feed-stocks have proved their potentiality towards the biomethane or biogas production both as single and cosubstrate. However, for the feedstocks with higher lignin content (straw, husk, stalk, etc.), an alkaline or acid pretreatment may impact the yield of biogas or biomethane (Shetty et al., 2017; Ward-Doria et al., 2016). In this context, the concept of energy crops is more extended present as a new class of feedstocks. Some instances of energy crops to produce biomethane has been discussed by several research groups, who took maize, miscanthus, switchgrass, hemp, sorghum, water hyacinth and several perennial crops as the substrate in both single and co-digestion process (Mayer et al., 2014; Priya et al., 2018; Schmidt et al., 2018). Feedstock plays a vital role in the biological methane production on which the process is dependent directly.

4.4 PRETREATMENT TECHNIQUES OF LIGNOCELLULOSIC BIOMASS

The pretreatment techniques are executed to reduce the crystallinity of cellulose and breakdown the barriers of recalcitrant lignin structure to improve the accessible surface area required for hydrolysis of sugars and enzymatic reactions (Koupaie et al., 2019; Kim et al., 2018). The ideal pretreatment technique offers simplicity with the effectiveness of removing lignin without degrading the sugars and increase their digest-ibility and yield through the AD process (Pellera and Gidarakos 2018; Venturin et al., 2018). Different pretreatment techniques are available that are applied to lignocellulose to solve the limited large retention time and improve the process of hydrolysis, enhance the efficacy of the generation of bioenergy and energy recovery as shown in Figure 4.3. For deconstructing lignin structure, the physical, chemical, thermochemical, biological, and hybrid techniques are available (Bhatt and Shilpa, 2014: Li et al., 2016).

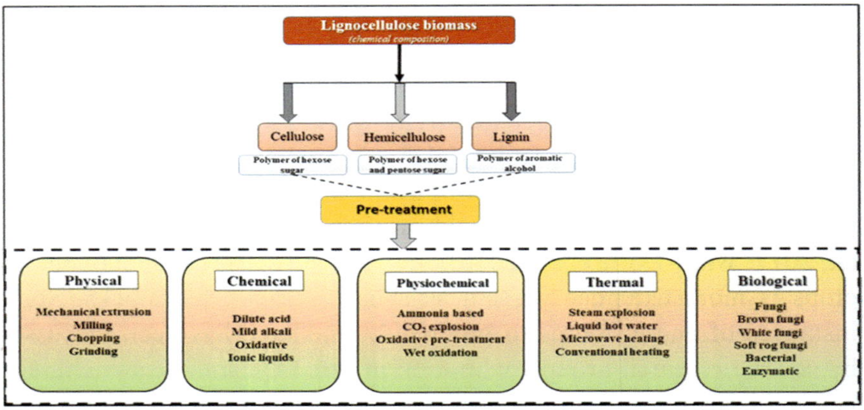

FIGURE 4.3 Different pretreatment techniques available for lignocellulose biomass.

Physical methods include reduction of particle size of the organic-rich substrate either by chopping, milling, or grinding. The biological method-based pretreatments use diverse classes of fungi such as white, brown or soft rogue types, enzymes, and bacteria (Sołowski et al., 2020) for degrading the complex carbohydrate structure as mentioned above (Ali et al., 2017). The chemical methods include treating the substrate with acidic, alkaline, oxidative or ionic liquid reacting agents (Sarto et al., 2019). A competitive summary of different pretreatment techniques of lignocellulose biomass is provided in Table 4.1. Maximizing the biogas production to alleviate the generation of a clean source of energy, is a key step in contributing towards a viable economy and sustainability of generating biogas through a renewable and alternative source like lignocellulosic biomass.

4.5 CHEMICAL PRETREATMENT OF LIGNOCELLULOSIC BIOMASS

Chemical pretreatment technology is a common and promising pretreatment technique applied to improve the digestibility of the treated substrate with minimum impacts on the environment (Tayyab et al., 2018). By applying alkaline pretreatment (e.g., KOH, NaOH, $Ca(OH)_2$ or aqueous NH_3), the intermolecular ester bonds between lignin from hemicellulose break effectively owing to the weak hydrogen bonds between cellulose and hemicellulose structure (Gu et al., 2015; Yadav et al., 2019). It has the upper hand over acids (inorganic and organic) such as hydrochloric acid, sulfuric acid, phosphoric acid or acetic acid in terms of generating corrosiveness in the reactor (Timung et al., 2015; Kim et al., 2018; Syaichurrozi et al., 2019).

Oxidative treatment is executed with peroxides, with their majority concentrating on hydrogen peroxide or ozonation or organic solvents reagents and any other pretreatment methods like enzymatic hydrolysis (Koupaie et al., 2019; Tapadia-Maheshwari et al., 2019; Pellera and Gidarakos 2018). In this review, we focus on the impact of chemical pretreatment methods using different chemical reagents as it has a decisive role in the solubilization and degradation of the complex-structured substrate. Different classes of chemical agents explored for chemical pretreatment is presented in a summarized form in Table 4.2.

4.5.1 ALKALI PRETREATMENT

Using alkaline agents like KOH or NaOH could help dismantle the crystalline structure of cellulose which supports hydrolysis. Delignification and reduction of cellulose density could be achieved if lignocellulose rich biomass is treated with alkaline agents before anaerobic digestion. Moreover, alkaline agents are used for some specific lignocellulosic biomass which does not degrade using acids.

Shen et al. (2019) investigated the impact of both mono-pretreatment and copretreatment techniques of varying concentrations for wheat straw as the starting material using KOH and $Ca(OH)_2$. They observed that the performance of methane yield via copretreatment technique with the combination of 2% KOH combined with 1% $Ca(OH)_2$ concentration displayed comparable results than the mono-treatment technique using only 3% KOH concentration. The cumulative methane yield from the copretreatment technique was 239.8 mL/g VS and biodegradability improved from 56% of raw wheat straw to 66%. Co-pretreatment technique using the mixture of KOH and $Ca(OH)_2$ was the optimal pretreatment strategy investigated using raw wheat straw.

Sharma and Singhal (2016) studied the influence of both acid (H_2SO_4) and alkaline (NaOH) medium for delignification of wheat straw (WS) in terms of biogas generation (mL/g VS) and enhancement of methane yield (%). They compared both the biogas generation (mL/g VS) and methane yield (%) from untreated WS and pretreated and methane content of 64%,

TABLE 4.1 Various Pretreatment Techniques for Lignocellulosic Rich Organic Substrate Before Anaerobic Digestion

Pretreatment techniques	Function	Challenges	References
Physical • Mechanical extrusion • Milling • Chopping • Grinding	• The reduction of particle size to fine powder contributes to increased crystallinity. • Operation time is less with a high yield of sugar. • A cost-effective technique for herbaceous and agriculture residue. • No inhibitory compound generation.	• Energy demand is high especially for hardwood biomass and maintaining high pressure is a challenge. • Not sustainable for the industrial-scale set up because of the high consumption of power. • Need special design of equipment and design.	Lindner et al. (2015); Krishania et al. (2013)
Chemical • Acid hydrolysis • Alkaline hydrolysis • Ionic liquids • Oxidative	• An effective technique in terms of partial to complete delignification. • Reduced cellulose crystallinity and higher solubilization of hemicellulose with the increased accessible surface area. • Short reaction and residence time. • Chemical pretreatment is generally associated with the highest pretreatment rate.	• Excessive consumption of chemicals is not an environmentally feasible approach along with the high cost of the chemicals. • Corrosion to reaction equipment with prolonged exposure. • Inhibitors and degradation products are produced.	Ali et al. (2020); Syaichurrozi et al. (2019); Sharma and Singhal (2016); Yadav et al. (2019)
Physiochemical • Ammonia fiber expansion • (AFEX) • Extrusion combined with chemical treatment	• Short retention time with the increased accessible surface area. • Lignin is dislocated from the cell wall and alters the structure depending on the treatment condition. • High digestibility for herbaceous and agriculture residues. • Hydrosylates are highly fermentable with lower inhibition.	• For AFEX ammonia additional recycling system is required. High cost associated with a large amount of NH_3. • For precursors like hardwood digestibility reported is poor and not effective for biomass with such high lignin concentration. • Minor effects on removed hemicellulose.	Gu et al. (2015)

TABLE 4.1 *(Continued)*

Pretreatment techniques	Function	Challenges	References
Thermal • Steam explosion • Microwave heating	• Partial to complete removal of hemicellulose. • Cost-effective technique. • The better extent of delignification.	• The high energy input of the thermal method is the major barrier against its industrial application. • The generation of toxic compounds makes the technique environmentally unfavorable.	Pellera and Gidarakos (2018); Lima et al. (2018); Theuretzbacher et al. (2015)
Biological • White rot fungi • Soft rot fungi	• No chemical products involved along with extra energy input makes the technically feasible and cost-effective. • They are known as the environmentally friendly and sustainable pretreatment strategy.	• Sometimes the rate of hydrolysis falls, impacting the overall efficacy of biogas generation. • The biological methods generally suffer from high costs, limiting their commercial application.	Shetty et al. (2020); Liu et al. (2017); Yadav et al. (2019); Zhang et al. (2016); Ghosh and Bhattacharyya (1999)

TABLE 4.2 Various Chemical Pretreatment Agents for Lignocellulosic Organic Substrate

Method	Agents	Features	References
Acid pretreatment	Sulfuric acid	Relatively cheap price for moderate material and efficient.	Syaichurrozi et al. (2019); Sharma and Singhal (2016)
	Phosphoric acid	Degradation of cellulose to volatile acids; small pH change	
	Nitric acid	High removal of hemicellulose; high methane production; expensive	
	Hydrochloric acid	Comparatively weak acid.	
Alkaline pretreatment	Sodium hydroxide	Decrease crystallinity of lignocellulose; cheap; increases corrosion risk; less efficient than KOH	Shetty et al. (2017)
	Potassium hydroxide	Efficient and expensive	
	Calcium hydroxide	Cheap and less efficient than NaOH and KOH	
Oxidation	Hydrogen peroxide	Hydrogen bonding in lignocellulose is cleaved.	Venturin et al. (2019)
	Peracetic acid	Relatively cheap	
	NaOH/urea	Relatively cheap	
Ionic liquid pretreatment	[BMIM] [CI]	Separates cellulose from lignin. Usually very costly.	Ali et al. (2020)
	EMIM] [AC]		
	$ZnCl_2$/HCl/HCHO		
	[BMIM]HSO_4		

whereas 2% H_2SO_4 gave the highest biogas yield of 140 mL/g. However, the highest methane content of 75% was obtained at the 5% acid concentration, respectively. Biogas yield of 128 mL/g and methane content of 71% were obtained at 2% and 5% NaOH concentrations, respectively. The findings suggested that pretreatment with 2% H_2SO_4 improved the performance of the bioenergy generation but with 5% H_2SO_4 more acid-soluble lignin was obtained. Moreover, the methane content in biogas was seen to be 10% more for WS treated with concentrated H_2SO_4 than NaOH. They suggested that acid-treated WS provide a pathway for the utilization of WS for energy production.

Patowary and Baruah (2018) studied the combined effect of thermal and alkali pretreatment of delignification on rice straw (RS) and corn stalk (CS). The potassium-based alkali reagent was obtained from burring banana peel ash, which is a rich source of alkali. $Ca(OH)_2$ is also considered as one of the most potential alkaline pretreatment agents for structural degradation of recalcitrant lignin. They observed that the chemical structure decomposition and biogas generation enhanced with the severity of the process conditions (temperature and duration). Lignin content reduced drastically for both RS and CS at severe process conditions, i.e., at higher temperature and duration of pretreatment. For chemically treated RS lignin content is reduced to 11.9 wt.% from 19.4 wt.% and for CS is reduced to 11 wt.% from 20.1 wt.% at 90°C. The biogas generation improved by 66% and 62% for pretreated RS and CS at 90°C for 10 h. The methane content in the biogas also improved, and the highest value obtained was on the 30th day for both RS and CS.

Zhang et al. (2015) reported the combined impact of extrusion and alkaline pretreatment using NaOH pretreatment on biogas generation using rice straw (RS). Firstly, the particle size of the biomass was reduced using an extruder before implementing the NaOH treatment on RS. 54% higher methane production was reported in their study at 3% of chemical loading along with 59.9% of improved energy recovery.

Yu et al. (2019) pointed out that in the case of anaerobic digestion using NaOH could cause a series of environmental impacts owing to the disposal of Na^+ ion in the black colored liquid phase than KOH in which the release of K^+ ion in the liquid could further be exploited as organic fertilizer. However, the main constraint using KOH as one of the effective alkaline agents lies in the fact that it is costly and could add extra cost to the entire system, which would not be an economically viable practice.

On the other hand, $Ca(OH)_2$ is a weak alkaline agent for anaerobic digestion of lignocellulose waste but in this regard, Ca $(OH)_2$ is comparatively cheaper and retainment of it in the anaerobic digestate will not bring any detrimental effect environmentally as Ca^{2+} has no inhibitory effect on anaerobic digestion. The use of $Ca(OH)_2$ would not bring disparaging impact both environmentally and economically.

Gu et al. (2015) investigated the combined effect of physical and chemical pretreatment on rice straw. They reported that varying concentrations of alkaline loading on extruded rice straw using $Ca(OH)_2$ improved the biogas production and enzyme hydrolysis. The chemical and physical structure of RS changed along with the reduction of fermentable sugar content. The highest biogas production reported by them is 574.5 mL/g VS at 10% of $Ca(OH)_2$, which is 36.7% times higher than the un-treated RS. Although, the optimal condition to obtain high biogas generation with effective hydrolysis reported by them is at 8% loading of $Ca(OH)_{2.}$ On the other hand, combined hydrogen peroxide pretreatment in alkaline conditions has drawn considerable attention owing to maximum sugar yield, cost-effectiveness, and feasibility. Mild alkali pretreatment of biomass before anaerobic digestion could be an environmentally friendly approach as it is less harsh and can also be cost-effective and carried with efficacy in terms of biogas generation and methane yield at ambient conditions (Kumari and Das, 2015).

4.5.2 ACID PRETREATMENT

The conventional method of lignin hydrolysis and formation of hydrolysis products include a direct hydrolysis strategy using inorganic acid. For the pretreatment of lignocellulosic rich biomass substrate, a different form of acids both organic or inorganic such as nitric acid, hydrochloric acid, sulfuric acid, phosphoric acid, acetic acid, maleic acid, citric acid, and lactic acid in concentrated or diluted form has been investigated a lot and have already been explored on the industrial scale also (Krishania et al., 2012; Solarte-Toro et al., 2019). Hemicellulose degraded to xylan is stable in acid solution. The acidic pretreatment is the most used technique due to its effectiveness and low cost. The usage of acid mainly concentrated ones could be the reason for the corrosive activity of the material. The acid pretreatment technique could become an economically feasible technique if after hydrolysis, the acid could be recovered and reused in a circular pattern.

Sarto et al. (2019) investigated the impact of sulfuric acid pretreatment technique and operating parameters like residence time under the batch system and ambient temperature and pressure conditions on biogas generation from water hyacinth. They varied two process parameters to optimize the biogas generation using water hyacinth. They reported delignification, degradation of cellulose to simple organic compounds and reduction of glucose, chemical oxygen demand (COD) and COD/nitrogen ratio with the increment of sulfuric acid concentration in water hyacinth. They reported the highest biogas obtained was at 5% loading of sulfuric acid and for 60 min of retention time. The highest value of biogas generated using water hyacinth was more than 420 mL with the highest methane content of approximately 64%. An enhancement of more than 130% of the biogas generation was found by them. A similar finding was reported by (Sharma and Singhal, 2016) who obtained the highest biogas yield using the acid pretreatment compared to the alkali pretreatment. The highest biogas generation reported was 140 mL with 72% improvement in methane content at 2% of acid loading.

Using sulfuric acid as the pretreatment agent for lignocellulose-based biomass is one of the conventional techniques because of its low cost and availability to disrupt the recalcitrant lignin structure and susceptibility of cellulose to enzymatic hydrolysis. Although, its use as an acidic pretreatment agent is limited due to the extent of corrosive activity and formation of inhibitors. The toxicity and corrosiveness generated using acid could be controlled by using a reactor made up of appropriate material which could sustain it for a prolonged time, making the entire system economically viable (Syaichurrozi et al., 2019; Krishania et al., 2012).

4.5.3 IONIC LIQUID PRETREATMENT

Owing to the strong hydrogen bond accepting tendency of the ionic liquid fractures, it augments the degradation of cellulose by ripping-off the cellulosic crystalline structures and eliminates traces of lignin (Ali et al., 2020). As compared to dilute acid pretreatment, ionic liquid pretreatment enhances the enzymatic hydrolysis (Li et al., 2010). Advantages of using ionic liquid pretreatment include appropriate chemical stability application and application over a large stable temperature range (Ouellet et al., 2011).

However, the limitation of using ionic-liquid pretreatment lies in its high cost and toxicity to microorganisms and enzymes.

Ali et al. (2020) investigated the combined effect of H_2O_2 as the alkaline agent and lithium chloride/N,N-dimethylacetamide (AHP-LiCl/DMAc) as the ionic-liquid on the accumulative methane generation from corn stalk. AHP-LiCl/DMAc is an inexpensive solvent system for deconstructing cellulose beyond derivatization (Alexandridis et al., 2018). They reported that the biomethane potential of CS was enhanced and AHP-LiCl/DMAc is a promising pretreatment medium that could be further explored on an industrial scale. It enhanced the accumulative methane yield and it is attributed to the delignification and enhanced cellulose disbanding of CS by adding ionic solvent of LiCl/DMAc. They reported improved biomethane yield over pristine CS and AHP-treated CS by 40% and 10%, respectively. The marginal difference of methane yield for both treated groups was substantial. The highest methane yield obtained was 318.6±17.85 mL/g VS for AHP-LiCl/DMAc pretreatment. Qi et al. (2017) also explored the use of LiCl/DMAc with acid for wheat stalks (WS) and reported an excessive loss of lignin.

4.5.4 OXIDATION

Venturin et al. (2018) studied the influence of H_2SO_4 and H_2O_2 on the co-digestion of swine manure and corn peduncle. They observed that H_2O_2 is an effective pretreatment agent that removed a considerable amount of lignin (71.6%) followed by hemicellulose (19.3%) and improved the cellulose content by 73.4%. Moreover, by using H_2O_2, the final volume of biogas increased compared to H_2SO_4. Their findings suggest that H_2O_2 could be effective as well as a cheap alternative reactive agent that could be further explored. Chemical pretreatment technology is a common and promising pretreatment technique applied to improve the digestibility of the substrate with minimum impacts on the environment (Tayyab et al., 2018). Chemical pretreatment such as alkaline pretreatment, using KOH, NaOH, $Ca(OH)_2$ or aqueous NH_3, helps to decline the hydrogen bond strength between cellulose and hemicellulose and break the intermolecular ester bonds between lignin from hemicellulose effectively (Gu et al., 2015; Yadav et al., 2019).

4.6 EFFECTS OF PROCESS PARAMETERS ON ANAEROBIC DIGESTION

In the biomethanation process, lignocellulosic biomass is valorized into energy carriers such as biogas, biomethane, and enriched manure. The entire process of biogas generation is executed inside the anaerobic digester and reactor with the help of microorganisms in the absence of oxygen, and their growth rests on different operating parameters like temperature, pH, organic loading, hydraulic retention time (HRT), carbon-to-nitrogen ratio (C/N) and reactor design of the anaerobic digester (Krishania et al., 2013; Li et al., 2020). Therefore, to exploit the potential of anaerobic digestion to the greatest extent in terms of quality and quantity of biogas generation, these parameters should be optimized and is represented schematically in Figure 4.4.

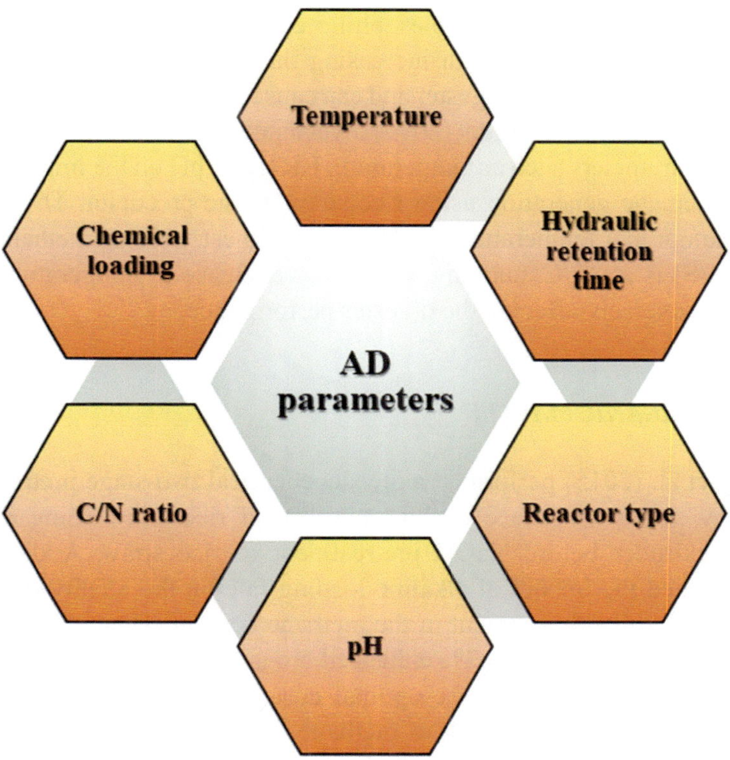

FIGURE 4.4 Different parameters influencing anaerobic digestion process.

4.6.1 TEMPERATURE

Temperature is one of the crucial parameters that determine the efficacy of an anaerobic digester. It has been reported that the temperature condition for thermophilic is in the range of 50–60°C followed by mesophilic in the range of 30–40°C, and psychrophilic under 20°C. The activity of anaerobic bacteria is affected by a slight variation in the temperature range inside the digester, dropping the rate of biogas generation (Mishra et al., 2018).

Typically, mesophilic bacteria can tolerate up to ±3°C of temperature fluctuation inside the digester. Moreover, providing a constant temperature is essential to avoid any negative impact on bacterial degradation and biogas generation. The microorganisms grow faster at higher temperature conditions than under mesophilic conditions reducing the HRT. For example, under thermophilic conditions, wood chips degrade completely into simple organic compounds in 11 days, whereas it took 20 days for complete degradation under mesophilic conditions (Hegde and Pullammanappallil, 2007). Although increasing the temperature could increase the imbalance inside the digester and can cause ammonia toxicity. Tapadia-Maheshwari and co-workers (2019) investigated the influence of different process parameters such as temperature, HRT, and pH on the improvement of biomethane generation using rice straw as the precursor. They found that the optimal temperature to obtain the highest yield of methane (274 mL/g VS) is at 37°C, and any increase or decrease of temperature 37°C beyond adversely affected the digester performance.

4.6.2 ORGANIC MATTER

Zhang et al. (2015) performed a physicochemical two-stage pretreatment strategy, firstly to reduce the particle size of rice straw using a twin-screw extruder before NaOH pretreatment of rice straw. Additionally, they studied the impact of alkaline loading rate on the quality of biogas production and enhancement in the methane content. They observed that an alkaline loading rate at 3% enhanced biogas production to the greatest extent. They reported that the optimal dosage for an ALR is 3% which generated the highest amount of methane from the pretreated rice straw. ALR pretreatment improved the hydrolysis rate of the anaerobic digestion attributed to the delignification via partially opening the lignin valve. A similar observation was reported by Pang et al. (2008) where 6% ALR was reported to be the optimal dosage among the varying concentration of NaOH ranging from 4–10% for higher and better biogas yield for anaerobic digestion using corn stover as the starting material.

Syaichurrozi et al. (2019) studied the influence of sulfuric acid loading on biogas yield from *Salvinia molesta* carried under the batch system and ambient temperature and pressure conditions. The pretreatment decreased the lignin content, accelerating the rate of hydrolysis and enhanced biogas generation. Furthermore, cumulative biogas yield from pretreated *Salvinia molesta* using sulfuric acid of 2–6% concentration improved to 22.7–24.1 mL/g VS compared to the one obtained from pristine biomass 13.28 mL/g VS.

4.6.3 TYPE OF BIOREACTOR OR DIGESTER

There are different configurations of digester depending on the approach of anaerobic digestion, for instance; one and two-stage digesters, batch or continuous digesters, dry or wet digesters or digesters with a combination of different approaches. The type of reactor is highly influenced by the total solid (TS) content such as a continuous flow stirred tank reactor (CSTR) is explored when the content of solid substrates is high, while up-flow sludge blanket is exploited when using high-rate biofilm systems (Kaparaju et al., 2009). Several other factors simultaneously inside the reactor influence the efficacy of biogas generation such as the pH level, temperature, and solid content.

4.6.4 HYDRAULIC RETENTION TIME

The number of days the waste is retained inside a tank where the anaerobic digestion is executed is defined as the hydraulic retention time (HRT). It is a significant parameter that influences the AD process directly as it creates the amount of time required for the development of microorganisms and their subsequent degradation to biogas and digester. It is measured by quantifying two parameters, i.e., the total volume of the tank and the daily flow. The influence of HRT is dependent on the temperature of the digester, type of substrate, and concentration of solids. HRT and the temperature inside the digester are inversely related, where with increasing temperature the HRT decreases (Singh and Maheshwari, 1995).

4.6.5 CARBON/NITROGEN RATIO

The carbon-to-nitrogen (C/N) is one of the significant parameters that influence digestion directly. Higher nitrogen content (low C/N ratio) in the substrate means undissociated ammonia, which could result in toxicity. On the other hand, a lower quantity of nitrogen (high C/N ratio) could inhibit the overall pace of anaerobic digestion. Maintaining an optimized range of the C/N ratio of the substrate is essential to improvise the generation, quality, and rate of digestion for the AD process. Several factors can contribute to maintaining a desired range of C/N ratio, such as the type of reactor used, co-digestion of different compatible substrates. Elsayed et al. (2016) investigated the influence of the C/N ratio on CH_4 yields. They reported highest CH_4 yields by increasing the C/N ratio to 10. Wang et al. (2012) also reported that better digestion performance could be obtained at a higher C/N ratio and stable pH. They obtained the highest methane yield from dairy manure, chicken manure and wheat straw at a higher C/N ratio of 25:1.

4.6.6 pH VALUE

An optimized range of pH is necessary to promote the growth of microbes as well as the generation of biogas (Weiland, 2003). pH requirement is different at different stages of anaerobic digestion. Hence, it is executed as a two-stage process where acetogenesis/methanogenesis is processed under neutral conditions (pH 7) and hydrolysis or acidogenesis is executed under acidic condition (pH range of 5.5–6.5) (Heo et al., 2003). Although maintaining a constant pH value at each stage is a challenge because, on the one hand, pH value increases by the accumulation of ammonia due to the degradation of protein and contrary due to the degradation of organic matter into volatile fatty acid, pH value decreases (Michael, 2003). For methanogenesis or acetogenesis pH value going below 6.6 is undesirable, which could have a detrimental impact on biogas production. To maintain a constant pH, calcium hydroxide could be injected inside the digester.

4.7 GLOBAL BIOMETHANE MARKET

According to the United Nations report (UN DESA 2014), the world urban population accounts for 54%, and by 2050 it is projected to increase up to 66%. The global urban population will reach 2.5 billion by 2050, of which 90% growth is anticipated to take place in developing countries like Africa and Asia. Countries like India, China, and Nigeria will experience the biggest boost in the urban population. Presently, the world produces 2.01 billion tons of MSW annually, which is expected o reach 3.40 billion tons by 2050. The average world waste generated per person on a single day is 0.74 kg (World Bank SWM, 2020). In most developing countries, the improper waste management system has resulted in serious environmental and health problems, which could impact the global economy (Srivastava et al., 2015; Ziraba et al., 2016; Khan et al., 2019).

The Global market is seeing biogas and its commercial application as an emerging opportunity towards profitable business and building a green world. The high-density energy and the enormous availability of cheap feedstocks have created a conducive scenario for business purposes. The global market will witness robust growth and drastic change in the biogas sector (Yousuf et al., 2016; Biofuels Digest, 2019; Schmid et al., 2019). The changing policies in emission regulation and shifting interest towards renewable energy will boost the biogas industry growth all over the globe. In most of the developing countries, biogas plants are installed mostly on a small scale or domestic basis and generated biogas is used as cooking fuel or sometimes for power generation. The replacement of conventional fuel with biogas can reduce firewood consumption and reduce the rate of deforestation. Asian countries like Pakistan, Bangladesh, Nepal, Thailand, and Vietnam several programs for supporting industrial and domestic-based biogas plants (Scarlat et al., 2018; Patinvoh and Taherzadeh, 2019).

China, a global leader in biogas plants had 100,000 modern plants and 43 million domestic-based biogas plants with an annual capacity of 15 billion m^3 (9 billion m^3 of biomethane). China's biogas growth is expected to cross 50 billion m^3 in the coming years (Scarlat et al., 2018). In Vietnam, there were around 158,500 domestic-based biogas digesters, which were accessed by nearly 79,000 rural population. The country is doing well in reducing CO_2 emissions up to 800,00 tons annually. The commendable achievements in the reduction of emissions have brought much international recognition at the global platform and bagged many awards for its tremendous performance in the biogas sector (Vietnam Biogas Program, 2020).

Bangladesh started the National Domestic Biogas and Manure Program for rural areas in the year 2006 and by 2014 installed 36,000 domestic-based biogas digesters and built around 500–600 commercial biogas plants. Nepal had started its biogas projects back in 1970 by installing 250 plants. Over the years, with the support of government and foreign aids, the biogas plant number has reached 300,000. In Nepal, the major hinder obstructing biogas plant success is the lack of availability of feedstocks and water, cold weather in hilly regions, and high installation cost (Patinvoh and Taherzadeh, 2019). Indonesia and Malaysia are leading palm oil producers, and the resulting effluent from these oil mills generates enormous waste, which makes it a huge potential for biogas production. Indonesia and Malaysia produce 4.35 billion m^3 and 3 billion m^3 of biomethane, respectively, from palm oil mill effluents. Pakistan started its biogas project in the year 2000, and currently, 1200 plants are functioning. Another 10,000 biogas plants are projected to install in the coming five years. The slow growth in biogas plants is a result of low-quality maintenance (BioEnergy Consult, 2019).

Africa, with a large amount of waste, has a huge potential for producing biomethane. Several countries like Ethiopia, Kenya, South Africa, Ghana, Guinea, Zimbabwe, Lesotho, Burundi, Nigeria, and Uganda have installed biogas plants and are supported under national programs (Austin and Morris, 2012; Kemausuor et al., 2018). Netherlands Development Organization supported the biogas partnership program in Africa has projected the installation of 100,000 domestic-based biogas digesters (Van Nes and Nhete, 2007). The initiative "Biogas for better life" is committed to enabling 2 million domestic biogas digesters by 2020, which will replace conventional cooking fuels. Around 18.5 million domestic biogas plants are estimated to be installed in Africa in the coming years (Africa Biogas Partnership Program, 2017).

Several European countries are now shifting interest in biomethantion technology and focused on reducing dependency on natural gas and other fossil fuels. The Biomethane process is extensively explored for its application as automobile fuel and injection into the natural gas grid (Grimalt-alemany and Engineering, 2017; Zhu et al., 2019).

In Latin America and Caribbean countries, numerous domestic biogas plants have been installed, which use agricultural, animal husbandry, and landfill wastes as feedstocks. Many programs are supporting small scale biogas digester installment in countries like Ecuador, Peru, Bolivia, Costa Rica, and Mexico. There are over 1000 household biogas digesters in Bolivia and big scale plants are being installed ion countries like Argentina, Columbia, and Honduras, which uses effluents generated from palm oil mill and other farms In Brazil, 127 biogas plants producing nearly 584 billion m^3 of biogas annually (Mittal et al., 2018).

The US has over 2200 biogas plants, of which 652 are using feedstocks from landfills, 250 are farm-based, 66 plants process food waste, and there are 1269 biogas digesters based on water resources facilities. There are 14958 new biogas plants projected to install in the coming years, which is estimated to produce 103 trillion kW of electricity. The produced energy can reduce a huge volume of emissions, which will be equivalent to 117 million vehicles removed from the road. The new development would result in creating huge job employment opportunities. The opportunities include 37,400 short-term employment in construction and 25,000 permanent jobs for operation and maintenance of the plant (American Biogas Council, 2020).

4.8 CONCLUSIONS

The high energy potential and capability in reducing greenhouse gases (GHGs) have made biomethanation popular among other renewable energy sources. The increasing concern over environmental degradation has pushed the world towards biomethanation, and with its social and economic importance, it has emerged as a suitable alternative to fossil fuels. Biomethanation has become the most economical and convenient method for the efficiency of the methanogens to produce methane from a wide range of feedstocks like municipal solid waste, agro-based residues, animal wastes, energy crops and various other lignocellulosic biomasses in a cost-effective way.

The production of biogas from the animal fecal has been upgraded to the more significant platform of producing biomethane from the above feedstocks with or without proper pretreatments. Besides the single feedstock, the co-digestion of lignocellulosic biomass with the municipal solid waste and animal wastes has been proved to be comparatively more potent to produce biomethane. Pretreatment is fundamental to enhance the rate of hydrolysis, which may be carried via different techniques. Chemical pretreatment strategies using different chemical agents and processes have gained considerable attention because of the enhanced biogas production and cumulative methane yield, feasibility for industrial-scale operation and low cost.

Suitable reactor design and large hydraulic retention time could further improve the quality of biogas generation and the quantity of methane yield. Further, optimizing other process parameters like reaction temperature, C/N ratio or pH could alleviate the quantity and quality of biogas generation to the maximum. The global biomethanation market is going to witness a major change in the developing economy, which is anticipated to reach an all-time high soon. Apart from addressing environmental mitigation, emerging biomethanation technology is projected to create many job opportunities all over the globe. Developing and developed countries are switching towards biomethanation and have increased the investment to encourage its wide use.

ACKNOWLEDGMENTS

The authors would like to thank the Natural Sciences and Engineering Research Council of Canada (NSERC) and Canada Research Chairs (CRC) program for funding this bioenergy research.

KEYWORDS

- **anaerobic digestion**
- **biofuel**
- **biogas**
- **biomethane**
- **chemical oxygen demand**
- **greenhouse gas**
- **lignocellulosic biomass**

REFERENCES

Adekunle, K. F., & Okolie, J. A., (2015). A review of biochemical process of anaerobic digestion. *Adv. Biosci. Biotechnol., 6*, 205–212.

Ali, N., Hamouda, H. I., Su, H., Li, F. L., & Lu, M., (2020). Combinations of alkaline hydrogen peroxide and lithium chloride/N, N-dimethylacetamide pretreatments of corn stalk for improved biomethanation. *Environ. Res., 186*.

Anderson, K., Sallis, P., & Uyanik, S., (2003). Anaerobic treatment processes. In: Mara, D., & Horan, N., (eds.), *Handbook of Water and Wastewater Microbiology* (pp. 391–426). Academic Press: London.

Angelidaki, I., Karakashev, D., Batstone, D. J., Plugge, C. M., & Stams, A. J., (2011). Biomethanation and its potential. In: Spada S., & Galluzzi L., (eds.), *Methods in Enzymology* (pp. 327–351). Academic Press.

Aziz, N. I. H. A., Hanafiah, M. M., & Ali, M. Y. M., (2019). Sustainable biogas production from agro-waste and effluents-A promising step for small-scale industry income. *Renew. Energy, 132*, 363–369.

Bhatt, S. M., & Shilpa, (2014). Lignocellulosic feedstock conversion, inhibitor detoxification and cellulosic hydrolysis: A review. *Biofuels, 5*(6), 633–649.

Bhutto, A. W., Qureshi, K., Harijan, K., Abro, R., Abbas, T., Bazmi, A. A., & Yu, G., (2017). Insight into progress in pretreatment of lignocellulosic biomass. *Energy, 122*, 724–745.

Chandra, R., Takeuchi, H., & Hasegawa, T., (2012). Methane production from lignocellulosic agricultural crop wastes: A review in context to second generation of biofuel production. *Renew. Sust. Energ. Rev., 16*, 1462–1476.

Cruz-Salomón, A., Cruz-Salomón, E., Pola-Albores, F., Lagunas-Rivera, S., Cruz-Rodríguez, R. I., Cruz-Salomón, K. D. C., & Domínguez-Espinosa, M. E., (2020). Treatment of cheese whey wastewater using an expanded granular sludge bed (EGSB) bioreactor with biomethane production. *Processes, 8*(8), 931.

Dollhofer, V., Dandikas, V., Dorn-In, S., Bauer, C., Lebuhn, M., & Bauer, J., (2018). Accelerated biogas production from lignocellulosic biomass after pretreatment with *Neocallimastix frontalis*. *Bioresour. Technol., 264*, 219–227.

El-Mashad, H. M., (2013). Kinetics of methane production from the codigestion of switchgrass and Spirulina platensis algae. *Bioresour Technol., 132*, 305–312.

Elsayed, M., Andres, Y., Blel, W., Gad, A., & Ahmed, A., (2016). Effect of VS organic loads and buckwheat husk on methane production by anaerobic codigestion of primary sludge and wheat straw. *Energy Convers Manag., 117*, 538–547.

Environmental and Energy Study Institute (EESI), (2017). *Fact Sheet-Biogas: Converting Waste to Energy*. https://www.eesi.org/papers/view/fact-sheet-biogasconverting-waste-to-energy (accessed on 25 June 2021).

Ghosh, A., & Bhattacharyya, B. C., (1999). Biomethanation of white rotted and brown rotted rice straw. *Bioprocess Eng., 20*(4), 297–302.

Grimalt-Alemany, A., Skiadas, I. V., & Gavala, H. N., (2018). Syngas biomethanation: State-of-the-art review and perspectives. *Biofuels Bioprod. Bioref., 12*(1), 139–158.

Gu, Y., Zhang, Y., & Zhou, X., (2015). Effect of $Ca(OH)_2$ pretreatment on extruded rice straw anaerobic digestion. *Bioresour. Technol., 196*, 116–122.

Hirunsupachote, S., & Chavalparit, O., (2019). Predicting the biomethanation potential of some lignocellulosic feedstocks using linear regression models: The effect of pretreatment. *KSCE J. Civ., 23*(4), 1501–1512.

Holmes, D. E., & Smith, J. A., (2016). Biologically produced methane as a renewable energy source. In: Sima, S., & Geoffrey M. G., (eds.), *Advances in Applied Microbiology* (Vol. 97, pp. 1–61). Academic Press: Cambridge.

Hosseini, K. E., Dahadha, S., Bazyar, L. A. A., Azizi, A., & Elbeshbishy, E., (2019). Enzymatic pretreatment of lignocellulosic biomass for enhanced biomethane production: A review. *J. Environ. Manag., 233*, 774–784.

Kemausuor, F., Adaramola, M. S., & Morken, J., (2018). A review of commercial biogas systems and lessons for Africa. *Energies, 11*(11), 2984.

Khan, B. A., Cheng, L., Khan, A. A., & Ahmed, H., (2019). Healthcare waste management in Asian developing countries: A mini-review. *Waste Manage. Res., 37*(9), 863–875.

Kim, M., Kim, B. C., Nam, K., & Choi, Y., (2018). Effect of pretreatment solutions and conditions on decomposition and anaerobic digestion of lignocellulosic biomass in rice straw. *Biochem. Eng. J., 140*, 108–114.

Krishania, M., Kumar, V., Vijay, V. K., & Malik, A., (2012). Opportunities for improvement of process technology for biomethanation processes. *Green Processing Synth., 1*(1), 49–59.

Krishania, M., Kumar, V., Vijay, V. K., & Malik, A., (2013). Analysis of different techniques used for improvement of biomethanation process: A review. *Fuel, 106*, 1–9.

Kumari, S., & Das, D., (2015). Improvement of gaseous energy recovery from sugarcane bagasse by dark fermentation followed by biomethanation process. *Bioresour. Technol., 194*, 354–363.

Li, M., Pu, Y., & Ragauskas, A. J., (2016). Current understanding of the correlation of lignin structure with biomass recalcitrance. *Front. Chem., 4*, 45.

Li, Y., Wang, Z., He, Z., Luo, S., Su, D., Jiang, H., & Xu, Q., (2020). Effects of temperature, hydrogen/carbon monoxide ratio and trace element addition on methane production performance from syngas biomethanation. *Bioresour. Technol., 295*, 122296.

Lima, D. R. S., Adarme, O. F. H., Baêta, B. E. L., Gurgel, L. V. A., & De Aquino, S. F., (2018). Influence of different thermal pretreatments and inoculum selection on the biomethanation of sugarcane bagasse by solid-state anaerobic digestion: A kinetic analysis. *Ind. Crops Prod., 111*, 684–693.

Lindner, J., Zielonka, S., Oechsner, H., & Lemmer, A., (2015). Effects of mechanical treatment of digestate after anaerobic digestion on the degree of degradation. *Bioresour. Technol., 178*, 194–200.

Liu, X., Hiligsmann, S., Gourdon, R., & Bayard, R., (2017). Anaerobic digestion of lignocellulosic biomasses pretreated with *Ceriporiopsis subvermispora. J. Environ. Manag., 193*, 154–162.

Lizasoain, J., Trulea, A., Gittinger, J., Kral, I., Piringer, G., Schedl, A., & Bauer, A., (2017). Corn stover for biogas production: Effect of steam explosion pretreatment on the gas yields and on the biodegradation kinetics of the primary structural compounds. *Bioresour. Technol., 244*, 949–956.

Luz, F. C., Cordiner, S., Manni, A., Mulone, V., & Rocco, V., (2017). Analysis of residual biomass fast pyrolysis at laboratory scale: Experimental and numerical evaluation of spent coffee powders energy content. *Energy Proc., 105*, 817–822.

Mayer, F., Gerin, P. A., Noo, A., Lemaigre, S., Stilmant, D., Schmit, T., & Foucart, G., (2014). Assessment of energy crops alternative to maize for biogas production in the greater region. *Bioresource Technol., 166*, 358–367.

Mishra, S., Singh, P. K., Dash, S., & Pattnaik, R., (2018). Microbial pretreatment of lignocellulosic biomass for enhanced biomethanation and waste management. *3 Biotech, 8*(11), 1–12.

Mittal, S., Ahlgren, E. O., & Shukla, P. R., (2018). Barriers to biogas dissemination in India: A review. *Energy Policy, 112*, 361–370.

Moghaddam, E. A., Ericsson, N., Hansson, P. A., & Nordberg, Å., (2019). Exploring the potential for biomethane production by willow pyrolysis using life cycle assessment methodology. *Energy Sustain. Soc., 9*(1), 1–18.

Molino, A., Nanna, F., Ding, Y., Bikson, B., & Braccio, G., (2013). Biomethane production by anaerobic digestion of organic waste. *Fuel, 103*, 1003–1009.

Mukherjee, A., Okolie, J. A., Abdelrasoul, A., Niu, C., & Dalai, A. K., (2019). Review of postcombustion carbon dioxide capture technologies using activated carbon. *J. Environ. Sci., 83*, 46–63.

Nanda, S., & Berruti, F., (2021). A technical review of bioenergy and resource recovery from municipal solid waste. *J. Hazard. Mater., 403*, 123970.

Nanda, S., Azargohar, R., Dalai, A. K., & Kozinski, J. A., (2015). An assessment on the sustainability of lignocellulosic biomass for biorefining. *Renew. Sustain. Energy Rev., 50*, 925–941.

Negi, S., Dhar, H., Hussain, A., & Kumar, S., (2018). Biomethanation potential for codigestion of municipal solid waste and rice straw: A batch study. *Bioresour. Technol., 254*, 139–144.

Neshat, S. A., Mohammadi, M., Najafpour, G. D., & Lahijani, P., (2017). Anaerobic codigestion of animal manures and lignocellulosic residues as a potent approach for sustainable biogas production. *Renew. Sust. Energ. Rev., 79*, 308–322.

Ngumah, C., Ogbulie, J., Orji, J., Amadi, E., Nweke, C., & Allino, J., (2017). Optimizing biomethanation of a lignocellulosic biomass using indigenous microbial-cellulases systems. *Biotechnol. Acta., 98*(3), 245–255.

Oechsner, H., Khanal, S. K., & Taherzadeh, M., (2015). Advances in biogas research and application. *Bioresour. Technol., 178*, 177.

Ogunleye, O. O., Aworanti, O. A., Agarry, S. E., & Aremu, M. O., (2016). Enhancement of animal waste biomethanation using fruit waste as cosubstrate and chicken rumen as inoculums. *Energy Sourc. Part A, 38*(11), 1653–1660.

Okeh, O. C., Onwosi, C. O., & Odibo, F. J. C., (2014). Biogas production from rice husks generated from various rice mills in Ebonyi State, Nigeria. *Renew. Energy, 62*, 204–208.

Okolie, J. A., Nanda, S., Dalai, A. K., & Kozinski, J. A., (2021). Chemistry and specialty industrial applications of lignocellulosic biomass. *Waste Biomass Valor.*, 12, 2145–2169.

Okolie, J. A., Nanda, S., Dalai, A. K., Berruti, F., & Kozinski, J. A., (2020). A review on subcritical and supercritical water gasification of biogenic, polymeric and petroleum wastes to hydrogen-rich synthesis gas. *Renew. Sust. Energy Rev., 119*, 109546.

Patil, V. S., & Deshmukh, H. V., (2015). A review on codigestion of vegetable waste with organic wastes for energy generation. *Int. Res. J. Biol. Sci., 4*(6), 83–86.

Patinvoh, R. J., & Taherzadeh, M. J., (2019). Challenges of biogas implementation in developing countries. *Curr. Opin. Environ. Sci. Health., 12*, 30–37.

Patowary, D., & Baruah, D. C., (2017). Effect of combined chemical and thermal pretreatments on biogas production from lignocellulosic biomasses. *Ind. Crops Prod., 124*, 735–746.

Pattanaik, L., Pattnaik, F., Saxena, D. K., & Naik, S. N., (2019). Biofuels from agricultural wastes. In: Basile, A., & Dalena, F., (eds.), *Second and Third Generation of Feedstocks: The evolution of Biofuels* (p. 103–142). Elsevier: Amsterdam.

Pellera, F. M., & Gidarakos, E., (2018). Chemical pretreatment of lignocellulosic agro-industrial waste for methane production. *J. Waste Manag., 71*, 689–703.

Prakash, E. V., & Singh, L. P., (2013). Biomethanation of vegetable and fruit waste in codigestion process. *Int. J. Emerg. Technol. Adv. Eng., 3*(6), 493–495.

Priya, P., Nikhitha, S. O., Anand, C., Nath, R. D., & Krishnakumar, B., (2018). Biomethanation of water hyacinth biomass. *Bioresour. Technol., 255*, 288–292.

Raveendran, S., Parameswaran, B., Beevi, U. S., Abraham, A., Kuruvilla, M. A., Madhavan, A., & Pandey, A., (2018). Applications of microbial enzymes in the food industry. *Food Technol. Biotech., 56*(1), 16–30.

Rouches, E., Escudié, R., Latrille, E., & Carrère, H., (2019). Solid-state anaerobic digestion of wheat straw: Impact of S/I ratio and pilot-scale fungal pretreatment. *J. Waste Manag., 85*, 464–476.

Saini, J. K., Saini, R., & Tewari, L., (2015). Lignocellulosic agriculture wastes as biomass feedstocks for second-generation bioethanol production: Concepts and recent developments. *3 Biotech, 5*(4), 337–353.

Sarker, S., Lamb, J. J., Hjelme, D. R., & Lien, K. M., (2019). A review of the role of critical parameters in the design and operation of biogas production plants. *Appl. Sci., 9*: 1915.

Sarto, S., Hildayati, R., & Syaichurrozi, I., (2019). Effect of chemical pretreatment using sulfuric acid on biogas production from water hyacinth and kinetics. *Renew. Energy, 132*, 335–350.

Scarlat, N., Dallemand, J., & Fahl, F., (2018). Biogas: Developments and perspectives in Europe. *Renew. Energy, 129*, 457–472.

Schmid, C., Horschig, T., Pfeiffer, A., Szarka, N., & Thrän, D., (2019). Biogas upgrading: A review of national biomethane strategies and support policies in selected countries. *Energies, 12*(19), 3803.

Schmidt, A., Lemaigre, S., Delfosse, P., Francken-Welz, H. V., & Emmerling, C., (2018). Biochemical methane potential (BMP) of six perennial energy crops cultivated at three different locations in W-Germany. *Biomass Convers. Biorefin., 8*(4), 873–888.

Schmidt, J. E., & Ahring, B. K., (1993). Effects of hydrogen and formate on the degradation of propionate and butyrate in thermophilic granules from an upflow anaerobic sludge blanket reactor. *Appl. Environ. Microbiol., 59*(8), 2546–2551.

Sharma, R., & Singhal, S., (2016). Pretreatment impact on biomethanation of lignocellulosic waste. *Single Cell Biol., 05*(01).

Shen, J., Zheng, Q., Zhang, R., Chen, C., & Liu, G., (2019). Co-pretreatment of wheat straw by potassium hydroxide and calcium hydroxide: Methane production, economics, and energy potential analysis. *J. Environ. Manag., 236*, 720–726.

Shetty, D. J., Kshirsagar, P., Tapadia-Maheshwari, S., Lanjekar, V., Singh, S. K., & Dhakephalkar, P. K., (2017). Alkali pretreatment at ambient temperature: A promising method to enhance biomethanation of rice straw. *Bioresour. Technol., 226*, 80–88.

Shetty, D., Joshi, A., Dagar, S. S., Kshirsagar, P., & Dhakephalkar, P. K., (2020). Bioaugmentation of anaerobic fungus *Orpinomyces joyonii* boosts sustainable biomethanation of rice straw without pretreatment. *Biomass Bioenergy, 138*, 105546.

Sivagurunathan, P., Kumar, G., Mudhoo, A., Rene, E. R., Saratale, G. D., Kobayashi, T., & Kim, D. H., (2017). Fermentative hydrogen production using lignocellulose biomass: An overview of pretreatment methods, inhibitor effects and detoxification experiences. *Renew. Sustain. Energy Rev., 77*, 28–42.

Solarte-Toro, J. C., Romero-García, J. M., Martínez-Patiño, J. C., Ruiz-Ramos, E., Castro-Galiano, E., & Cardona-Alzate, C. A., (2019). Acid pretreatment of lignocellulosic biomass for energy vectors production: A review focused on operational conditions and techno-economic assessment for bioethanol production. *Renew. Sustain. Energy Rev., 107*, 587–601.

Sołowski, G., Konkol, I., & Cenian, A., (2020). Production of hydrogen and methane from lignocellulose waste by fermentation. A review of chemical pretreatment for enhancing the efficiency of the digestion process. *J. Clean. Prod., 267*,121721.

Srivastava, V., Ismail, S. A., Singh, P., & Singh, R. P., (2015). Urban solid waste management in the developing world with emphasis on India: Challenges and opportunities. *Rev. Environ. Sci. Biotechnol., 14*, 317–337.

Sukasem, N., & Prayoonkham, S., (2017). Biomethane recovery from fresh and dry water hyacinth anaerobic codigestion with pig dung, elephant dung and bat dung with different alkali pretreatments. *Energy Proc., 138*, 294–300.

Syaichurrozi, I., Villta, P. K., Nabilah, N., & Rusdi, R., (2019). Effect of sulfuric acid pretreatment on biogas production from *Salvinia molesta*. *J. Environ. Chem. Eng., 7*(1), 102857.

Tapadia-Maheshwari, S., Pore, S., Engineer, A., Shetty, D., Dagar, S. S., & Dhakephalkar, P. K., (2019). Illustration of the microbial community selected by optimized process and nutritional parameters resulting in enhanced biomethanation of rice straw without thermochemical pretreatment. *Bioresour. Technol., 289*, 121639.

Tayyab, M., Noman, A., Islam, W., Waheed, S., Arafat, Y., Ali, F., Zaynab, M., et al., (2019). Bioethanol production from lignocellulosic biomass by environment-friendly pretreatment methods: A review. *Appl. Ecol. Environ. Res., 16*(1), 225–249.

Theuretzbacher, F., Lizasoain, J., Lefever, C., Saylor, M. K., Enguidanos, R., Weran, N., Gronauer, A., & Bauer, A., (2015). Steam explosion pretreatment of wheat straw to improve methane yields: Investigation of the degradation kinetics of structural compounds during anaerobic digestion. *Bioresour. Technol., 179*, 299–305.

Thi, N. B. D., Kumar, G., & Lin, C. Y., (2015). An overview of food waste management in developing countries: Current status and future perspective. *J. Environ. Manag., 157*, 220–229.

Timung, R., Mohan, M., Chilukoti, B., Sasmal, S., Banerjee, T., & Goud, V. V., (2015). Optimization of dilute acid and hot water pretreatment of different lignocellulosic biomass: A comparative study. *Biomass Bioenergy, 81*, 9–18.

Venturin, B., Frumi, C. A., Scapini, T., Mulinari, J., Bonatto, C., Bazoti, S., Pereira, S. D., et al., (2018). Effect of pretreatments on corn stalk chemical properties for biogas production purposes. *Bioresour. Technol., 266*, 116–124.

Wang, X., Yang, G., Feng, Y., Ren, G., & Han, X., (2012). Optimizing feeding composition and carbon-nitrogen ratios for improved methane yield during anaerobic codigestion of dairy, chicken manure and wheat straw. *Bioresour. Technol., 120*, 78–83.

Ward-Doria, M., Arzuaga-Garrido, J., Ojeda, K. A., & Sanchez, E., (2016). Production of biogas from acid and alkaline pretreated cocoa pod husk (*Theobroma cacao L.*). *Int. J. Chemtech Res., 9*(11), 252–260.

Yadav, M., Paritosh, K., Pareek, N., & Vivekanand, V., (2019). Coupled treatment of lignocellulosic agricultural residues for augmented biomethanation. *J. Clean. Prod., 213*, 75–88.

Yangin-Gomec, C., & Ozturk, I., (2013). Effect of maize silage addition on biomethane recovery from mesophilic codigestion of chicken and cattle manure to suppress ammonia inhibition. *Energy Convers. Manage., 71*, 92–100.

Yousuf, A., Khan, M. R., Pirozzi, D., & Ab Wahid, Z., (2016). Financial sustainability of biogas technology: Barriers, opportunities, and solutions. *Energy Sourc. B., 11*(9), 841–848.

Yu, Q., Liu, R., Li, K., & Ma, R., (2019). A review of crop straw pretreatment methods for biogas production by anaerobic digestion in China. *Renew. Sustain. Energy Rev., 107*, 51–58.

Yu, Z., & Schanbacher, F. L., (2010). Production of methane biogas as fuel through anaerobic digestion. In: Singh, O. V., & Harvey, S. P., (eds.), *Sustainable Biotechnology: Sources of Renewable Energy* (p. 105–127). Springer: Dordrecht.

Zhang, H., Zhang, P., Ye, J., Wu, Y., Fang, W., Gou, X., & Zeng, G., (2016). Improvement of methane production from rice straw with rumen fluid pretreatment: A feasibility study. *Int. Biodeter. Biodegrad., 113*, 9–16.

Zhang, Y., Bi, P., Wang, J., Jiang, P., Wu, X., Xue, H., & Li, Q., (2015). Production of jet and diesel biofuels from renewable lignocellulosic biomass. *Appl. Energy, 150*, 128–137.

Zhang, Y., Chen, X., Gu, Y., & Zhou, X., (2015). A physicochemical method for increasing methane production from rice straw: Extrusion combined with alkali pretreatment. *Appl. Energy, 160*, 39–48.

Zhu, T., Curtis, J., & Clancy, M., (2019). Promoting agricultural biogas and biomethane production: Lessons from cross-country studies. *Renew. Sust. Energ. Rev., 114*, 109332.

Ziraba, A. K., Haregu, T. N., & Mberu, B., (2016). A review and framework for understanding the potential impact of poor solid waste management on health in developing countries. *Arch Public Health, 74*(1), 1–11.

CHAPTER 5

Recent Advancements in Thermochemical Biomethane Production

BISWA R. PATRA,[1] JUDE A. OKOLIE,[1] ALIVIA MUKHERJEE,[1] FALGUNI PATTNAIK,[2] SONIL NANDA,[1] AJAY K. DALAI,[1] JANUSZ A. KOZINSKI,[3] and PRAKASH K. SARANGI[4]

[1]*Department of Chemical and Biological Engineering, University of Saskatchewan, Saskatoon, Saskatchewan, Canada*
E-mail: ajay.dalai@usask.ca (Ajay K. Dalai)

[2]*Center for Rural Development and Technology, Indian Institute of Technology Delhi, New Delhi, India*

[3]*Faculty of Engineering, Lakehead University, Thunder Bay, Ontario, Canada*

[4]*Directorate of Research, Central Agricultural University, Imphal, Manipur, India*

ABSTRACT

The recent prodigious increase in the global population has unprecedentedly increased energy demands. To cope with these demands, the need of the hour is to focus on alternative renewable and environmentally benign energy sources to limit the dependency on fossil fuels. Biomethane has emerged as one of the best alternatives to fossil fuel with many advantages. This chapter describes the technologies available for biomethane production. Anaerobic digestion process, pyrolysis, and gasification are the most commonly available biomethane production technologies. A clear distinction between these technologies is outlined in this chapter. Several feedstocks for biomethane production are comprehensively reviewed. This chapter assesses some recent advancements in biomethane production along with its socio-economic impacts and applications. Furthermore, the chapter explores the global approach towards a 'biomethane economy' together with its potential.

5.1 INTRODUCTION

The elevating world population and economic growth have augmented the demand for alternative energy sources. Moreover, a significant share of global energy consumption is reliant on fossil fuels. Excessive consumption of the use of fossil fuels has adverse effects on the environment because of greenhouse gas and CO_2 emissions (Nanda et al., 2015; Okolie et al., 2020). Furthermore, fossil fuels are associated with price instability and depletion concerns. These issues have obligated the demand for renewable and sustainable sources of energy. Among all the renewable energy sources, biofuels from thermochemical and biological conversion routes are emerging as one of the promising alternatives to fossil fuels (Nanda et al., 2014, 2016c).

Biofuels are viewed as carbon-neutral alternatives to petroleum resources in the energy and transportation industry (Nanda et al., 2017). Feedstocks for the production of biofuels are cheap, readily available and renewable. Biomethane is a versatile and easily stored biofuel that can be produced from biomass through a thermochemical process (i.e., gasification) or biological process (i.e., anaerobic digestion). Furthermore, biomethane is widely perceived to be more flexible in its utilization when compared with other renewable energy sources (Figure 5.1). Thus, its flexible and varied spectrum of applications in the field of production of electricity, heat provision and mobility has created an extensive base of potential consumers.

The anaerobic digestion process is the most popular route for biomethane production. The process, regarded as a biological conversion process, results in the conversion of organic materials into biogas, which is later upgraded to biomethane. The anaerobic process is a natural process, which occurs in landfills, wetlands, soils, gastrointestinal tracts of insects and animals in the

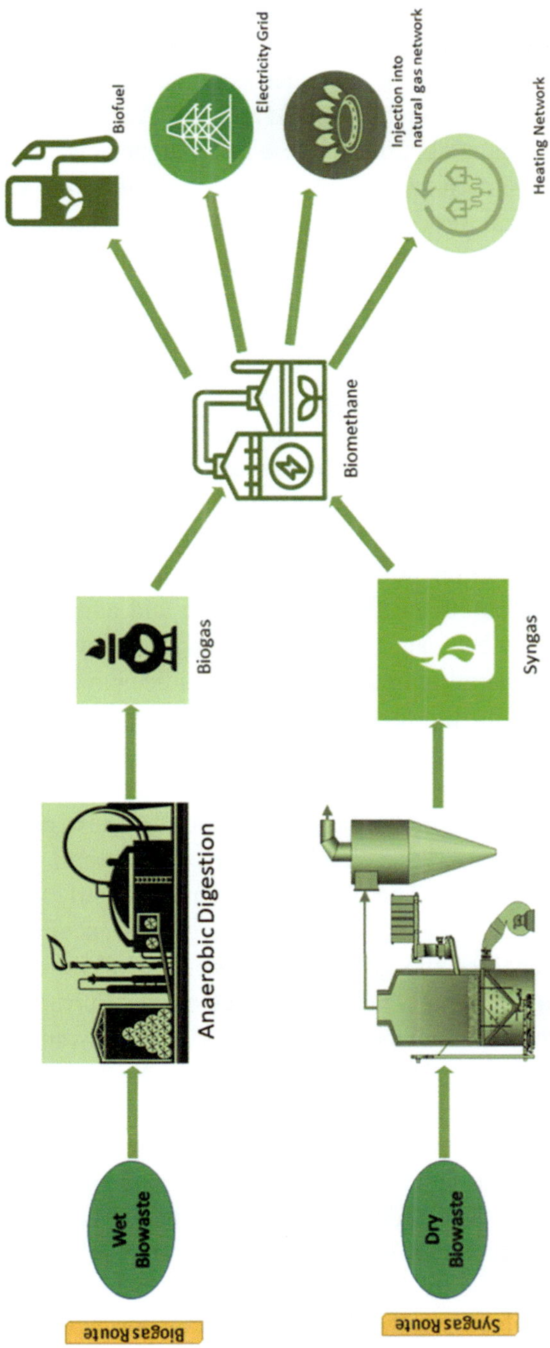

FIGURE 5.1 Different routes for biomethane production and its applications.

absence of light and oxygen, iron, nitrate like electron acceptor (Hattori, 2008). The anaerobic decomposition of organic matter occurs in the presence of three kinds of microorganisms such as fermentative bacteria, organic acid oxidizing bacteria and methanogenic archaea (Angelidaki et al., 2011; Pore et al., 2019; Saha et al., 2020). These microorganisms are responsible for producing different enzymes and metabolites, which are essential for continuing the overall process. Although anaerobic digestion has been successfully applied in the generation of biogas and subsequent upgrading to biomethane, some issues remain inherent. For example, the high cost of biogas upgrading technologies (e.g., adsorption process, scrubbing, etc.), the long residence time involved in the anaerobic digestion process and feedstock selectivity. These challenges have stimulated interest in thermochemical conversion routes such as gasification for biomethane production.

Several studies have compared the two technologies (i.e., gasification, and anaerobic digestion) for biomethane production (Li et al., 2017). Table 5.1 makes a comparison between biomethanation and gasification for biomethane production. A thermodynamic analysis that estimates the theoretic yield of biomethane from gasification and anaerobic digestion was comprehensively evaluated by Gallagher and Murphy (2013). Despite the vast amount of literature in this field, few studies have outlined recent advances in biomethane production. Therefore, this study aims to provide recent advances in this field. The thermochemical conversion process for biomethane production is discussed in this chapter, followed by recent research studies related to biomethane production by these processes.

Biomethane can be used as a direct alternative for natural gas in heating and electricity generation. Therefore, direct injection of biomethantion into the existing grid of natural gas has emerged as an efficient energy alternative. This enables the gas grid operators to allow consumers to make an easy transition to a renewable source of gas. Furthermore, it can be used for cooking, as a transportation fuel for light-duty and heavy-duty vehicles (Pääkkönen et al., 2019). Regarding the utilization of biomethane as gaseous fuels, some countries such as Sweden have already started exploring its usage as transport fuels. About 50% of stored biomethane are used as transportation fuels for vehicles in Sweden (Budzianowski and Brodacka, 2017). The reduced technological investment, efficiency, and sustainability approaches have made biomethanation a benign source for producing green energy, and the stored biomethane can be used for industrial heating, electricity generation and transport fuel. This enables the operators of the gas grid to let consumers in making a swift transition to a renewable and sustainable source of gas and the production of short-chain olefins (Chen et al., 2015).

TABLE 5.1 Comparison Between Anaerobic Digestion and Gasification for Biomethanation

Attribute	Anaerobic digestion	Gasification
Energy efficiency	The low energy requirement (heat and electricity) for the decomposition of complex organic matter makes it highly energy efficient.	The generation of biomethane from syngas is effective but requires an additional cleaning and methanation process to convert CO and H_2 to CH_4, which adds the requirement of excess energy. The lower overall efficiency of gasification.
Upgrading of biogas	Upgrading of biogas to biomethane is required to remove excess CO_2 to explore it as a vehicular fuel (equivalent to clean natural gas or compressed natural gas).	CO and H_2 upgraded to CH_4 via the methanation process after the water-gas shift process.
Overall cost	The overall cost of biomethane generation is high because of an additional step required for upgrading biogas to biomethane by pressure swing adsorption, water scrubbing or chemical absorption.	The methanation process to produce a high volume of CH_4 and the water-gas shift process makes the overall system configuration more complicated, higher investment and operation costs.

As stated earlier, there are two applied routes for biomethane production, i.e., thermochemical and biological routes. The latter involves the application of enzymes or microorganisms to decompose organic matter into green fuels and chemicals. Examples of such processes include fermentation and anaerobic digestion processes (Okolie et al., 2020). On the other hand, the thermochemical route applies heat and chemical energy to disintegrate organic compounds into solid, liquid, or gaseous fuels. Examples of thermochemical approaches include pyrolysis, conventional gasification, hydrothermal gasification and liquefaction. Concerning biomethane production anaerobic digestion, gasification, and pyrolysis are the proven technologies for its production, Although, some studies also propose an integrated system that comprises of two or three of the aforementioned processes (Lin et al., 2018; Luz et al., 2018; Lü et al., 2018).

5.2 THERMOCHEMICAL TECHNOLOGIES FOR BIOMETHANE PRODUCTION

Thermochemical technologies for producing biomethane integrates heat and chemical energy to convert biomass into useful products. Examples of such processes include conventional gasification, pyrolysis, and hydrothermal gasification (Okolie et al., 2019). The thermochemical route attempts to address issues with biological products such as the anaerobic digestion process. The anaerobic digestion process led biogas generation contains a large quantity of CO_2, which requires expensive upgrading to be converted to biomethane for transportation fuels (Li et al., 2017).

The gasification process involves a series of reactions in which carbonaceous materials are converted into synthetic gas or syngas in the presence of steam and/or oxygen. Examples of such reactions include the water gas shift reaction, combustion, methanation, and hydrogenation reactions as represented in Eq. (5.1–5.7) (Korres et al., 2010).

Combustion reactions:

$$C + O_2 \rightarrow CO_2 \tag{5.1}$$

$$C + 0.5O_2 \rightarrow CO \tag{5.2}$$

Water-gas shift reaction:

$$CO + H_2O \rightarrow CO_2 + H_2 \tag{5.3}$$

Hydrogenation reaction:

$$CO_2 + 2H_2 \rightarrow CH_4 + 1/2O_2 \tag{5.4}$$

Methanation reaction:

$$C + 2H_2 \rightarrow CH_4 \tag{5.5}$$

$$CO + 3H_2 \rightarrow CH_4 + H_2O \tag{5.6}$$

$$CO_2 + 4H_2 \rightarrow CH_4 + 2H_2O \tag{5.7}$$

Combustion reactions presented in Eq. (5.1) and Eq. (5.2) involve an exothermic reaction that produces heat for the subsequent methanation reactions, which are endothermic. During gasification, feedstocks such as woody biomass, food waste, and agricultural residues can be converted into syngas, which has diverse applications such as heat and power generation and direct vehicular fuel (Nanda et al., 2015). Syngas produced from gasification can be upgraded to produce biomethane. On the contrary, the subsequent upgrading to natural gas via the methanation process is not yet commercially demonstrated.

When biomass gasification occurs in hydrothermal conditions (i.e., temperature > 374.1°C and pressure > 22.1 MPa), the process is known as supercritical water gasification (Reddy et al., 2014, 2016; Nanda et al., 2016a). The latter has several advantages over conventional gasification, including rapid decomposition of feed materials, increased reaction rates even at lower temperatures, decrease in the production of chars and tars and rapid solubility of feedstocks in water (Okolie et al., 2019; Okolie et al., 2020).

Gasification is one of the most promising routes to produce biomethane from biomass materials. Compared to anaerobic digestion, the process requires a shorter residence time, offers high conversion efficiency and is compatible with a wide variety of feedstocks. Göteborg biomass gasification project (i.e., GoBiGas) is known as the first large-scale biomethane production plant in the world via the gasification process (Alamia et al., 2017). The plant, located in Sweden, uses forest residues as feedstock to produce biomethane in a fluidized-bed gasifier operating at 850°C by steam injection from a separate combustion chamber. Since then, there have been several researchers working on the direct production of biomethane from gasification.

Wang et al. (2015) performed the thermodynamic analysis of biomass gasification for biomethane production. The authors applied the Gibbs free minimization method to predict the theoretical methane yield from the gasification process. Their study showed that the addition of steam to the biomass gasification is conductive to augment the carbon conversion and results in a shrinking carbon deposit zone, but the use of other gasifying agents like air, O_2 and CO_2 will enforce a negative impact on the yield of methane and also affects the H_2/CO ratio. Gallagher et al. (2013) compared the gasification and anaerobic digestion process in Ireland in terms of net energy production. The authors noted that the net energy of both processes is similar.

Pyrolysis is also another thermochemical conversion process for biomethane production. The process involves thermal disintegration of materials in an inert environment to produce liquid, gaseous, and solid products. There are different forms of pyrolysis such as slow, fast, vacuum, and flash pyrolysis depending on the operating conditions such as temperature, residence time, heating rate, reactor type, and nature of feedstock (Mohanty et al., 2013). The feedstock is heated under high temperatures. Bio-oil and gaseous products are favored during fast or flash pyrolysis at a temperature in the range of 400–800°C with 10–200°C/s as heating rate and 0.5–5 s as residence time (Nanda et al., 2015). On the other hand, slow pyrolysis with a lower heating rate promotes the generation of biochars.

Görling et al. (2013) proposed biomethane generation from the poly generation plant along with the generation of biochar and heat from fast pyrolysis of biomass. The authors used ASPEN Plus simulation program to evaluate the energy and material flow. It was observed that about 15.5 MW and 3.7 MW of biomethane and biochar were produced respectively from a total energy input of 23 MW raw biomass and 1.39 MW electricity. The energy performance was accessed along with the climate impact of anaerobic digestion and pyrolysis for biomethane production using a lifecycle assessment method by Moghaddam et al. (2019). The analysis includes the whole technical system starting from the production and processing of biomass to the generation of biomethane as the end product. Besides, the authors also studied its impact on climate when resulting biochar as a by-product used for soil conditioning and as an agent for carbon sequestration. Their results showed that the pyrolysis process showed a higher external energy ratio, which was effective in mitigating climatic challenges as compared to the anaerobic digestion process. The probable reasons for the effectiveness of the pyrolysis process and upgrading units were found to be the lower inputs of primary energy and the diminished methane loss.

An integrated biological and thermochemical route for biomethane production has been investigated by several authors. Salman et al. (2017) simulated a system that integrates the pyrolysis and anaerobic digestion process to enhance biomethane production from organic waste materials. Firstly, the waste undergoes pyrolysis to produce biochar, producer gas and bio-oil. The latter was added as an adsorbent to the digester for improving the biomethane content. The producer gas and bio-oil were reformed via methanation reaction to produce biomethane. Compared to the energy efficiency of the separate anaerobic digestion (i.e., 52% energy efficiency), the integrated process exhibited a superior energy efficiency of 67%. Fabbri and Torri (2016) provided an overview of a different integrated process for biomethane production.

5.3 STORAGE TECHNOLOGIES FOR BIOMETHANE

The integrated energy system of biomass-derived methane comprises three major parts such as methane production, storage, and distribution. Among these three, storage can be considered as the most crucial stage in the integrated energy system for maintaining the techno-economic parameters and safety in the whole methane production module. Methane, being an inflammable gas, should be stored in a sophisticated way so that safety is appropriately provided and the on-site usage or the off-site distribution will be convenient (Barik et al., 2013). Biomethane can be stored and transported either in compressed or liquefied form. Biomethane derived via anaerobic digestion could contain traces amounts of impurities like H_2S, CO_2 and water vapor (Budzianowski and Brodacka, 2017). These impurity gasses can create severe corrosion to the storage devices. Hence, these impurities should be removed to obtain pure methane.

Storage of the biogas or biomethane is equivalently vital as the feedstocks and the process phases contribute directly to the safety, efficacy, and transportation of the biomethane. In the context of the storage capacity, there are three types of storage systems available named low, medium, and high-pressure storage systems depending upon the storage pressure range. Among these three storage systems, the low-pressure storage system (i.e., pressure of 2 psi) is relatively less expensive than the other two others due to the higher handling and maintenance cost (Bioenergy Consult, 2020). The high-pressure storage system is operated at 2000–5000 psi and biomethane can be stored in the form of compressed biomethane or liquefied biomethane. Therefore, the handling and transportation of biomethane become more accessible with a proper selection of the storage system. Economically, a medium pressure storage system (2–200 psi) is also used with partial compression of the gas (Bioenergy Consult, 2020).

For the transportation of the small amount of gas, people often use smaller gas bags operating at a mild pressure level, and this is comparatively handy to use (Arrhenius et al., 2019). In the case of on-site storage, the low pressure and medium pressure storage systems are appropriate for the gaseous biomethane or biogas. However, for off-site storage of a larger volume of biomethane, the storage space can be compacted to a smaller cylindrical storage system by upgrading or compressing the biomethane to compressed biomethane. In another way, to decrease the volume and to impose greater safety, biomethane is liquefied into liquefied biomethane by condensing the gaseous methane (Krich et al., 2019). Both compressed biomethane and liquefied biomethane are usually stored in high-pressure storage systems majorly operated at the pressure between 2000 to 5000 psi. Moreover, in high-pressure systems, the impurities should be removed strictly because, at this high-pressure range, these impurities like H_2S can significantly corrode the storage tanks.

Besides the storage tanks, in a recent case scenario, several porous adsorbents are being used to adsorb the biomethane for future use, which mainly includes activated carbon, zeolite, and metal-organic frameworks (Feroldi et al., 2016; Wang et al., 2017; Li et al., 2019). Majorly this form of storage is used for low-pressure storage and activated carbon is a comparatively cheaper option to store biomass-derived methane or other gasses like CO_2 (Nanda et al., 2016a; Matos et al., 2017). The storage of methane into these porous materials can be explored broadly in the future for maintaining economical, safe, and convenient storage of this inflammable gas.

For the on-site storage of biomethane, usually, a spherical storage tank is used due to its lower surface area or more moderate land use for storing a high volume of biomethane (Krich et al., 2019). The major components of a storage tank are the materials used in the inner or outer membrane, which is fabricated by the materials like reinforced plastics, steel, and alloy steel (Krich et al., 2019). These types of materials enhance the safety aspects of the biomethane storage tanks. The internal atmosphere of the storage tanks is preserved by the airflow system, which plays an important role in maintaining such a higher volume of methane gas more safely.

5.4 BIOMETHANE AND ITS SOCIO-ECONOMIC IMPACTS

The depletion of fossil fuels is deepening day by day with the deterioration of the environment because of excessive greenhouse gas emissions and global warming. The increasing concern over environmental degradation has led the world to focus on significant research on biomethane production. Biomethanation, with its environmental remediation benefits and sustainable nature, is gaining contemplation as a promising alternative to fossil fuel. The high energy potential and capability in reducing greenhouse gases make biomethanation hopeful for future energy security (Balussou et al., 2012). Renewable energy from different sources like wind, solar, tidal, and geothermal may be expensive to harness, store, and use efficiently. In contrast, biomethanation has the advantage of being one of the most economical and environmentally friendly technologies to generate biofuels while boosting the rural economy. Despite all such benefits, biomethanation plants often face social opposition based on concerns over health and environmental issues. The fact cannot be denied that there is a high possibility for the escape of greenhouse gases (e.g., NO_X, CO_2, and CH_4) from the anaerobic digestor.

The benefits and prospects of biomethantion indicate its future contribution to accelerate socio-economic and techno-economic development in providing clean energy. Industries can become self-sustained and can reduce expenditure by using their effluents and treating them to generate electricity and heat. The solid product can be used as a biofertilizer for soil conditioning, which can make a significant contribution to food security. In many developing and developed countries, biomethane production has undergone exponential research and development (Ferella et al., 2019). In many developing countries and rural communities, mostly animal manure and food waste are used as feedstocks for biogas plants. The solid product (i.e., biosludge) obtained from anaerobic digestion can be used as an organic fertilizer in agriculture. The uncontrolled landfills and waste dumping lead to the generation of foul odor, emission of greenhouse gases and the spreading of diseases. Waste management by using the waste for energy application reduces environmental deterioration and improves the health and sanitation of society. The numerous benefits make it popular among other renewable sources of energy, and its growing demand shows its acceptance in society. Biomethane produced via anaerobic digestion can be used for injection into the electric grid, vehicle fuel, and the energy can be stored to meet the demand of the peak load.

The wastewater utilization in biomethane production through anaerobic digestion can reduce the pollution in natural water bodies. The growing demand for biomethanation has created a positive impact on the global economy market. The global market is going to experience a major boost in the biogas economy, which is anticipated to surpass US $110 billion by 2025 (Global Market Insights, 2020).

Biomethanation is a boost to economic development and creates job opportunities. The job opportunities start with the short-term employment in the construction of biogas plants, while plant operation, manufacturing unit for equipment and plant maintenance provides long-term employment opportunities. The generation of electricity and its storage also provides large employment. The waste management and collection of food and biogenic wastes also require a huge workforce. Small scale biomethanation plant in villages improves the quality of life in communities of farmers and enhance their sanitation. It also helps them economically by improving crop production.

5.5 INDIA AS A CASE STUDY

India is the second-most populous country after China, with nearly 1.4 billion people. The emerging and fastest-growing economy has shifted agriculture driven nations to urbanization and industrialization. India, with diverse geographic and climatic conditions, has varied food consumption and dietary pattern and generates different kinds of wastes. The increasing urbanization and altering lifestyles have led Indian urban cities to generate solid waste of about 90 million tons annually as by-products from domestic, agricultural, industrial, and other activities.

The open and uncontrolled dumping yards are used to dump about 70–90% of municipal solid waste. The uncontrolled disposal of waste leads to local sanitation problem and affect human health adversely by becoming breeding yard for diseases. It also causes a serious threat to the environment by emitting greenhouse gases, and the runoff water from dumping yards pollutes groundwater and other water bodies. To reduce the impact of such threats, India has moved towards using alternative methods for waste management. Among all such measures, investment in biogas and biomethane production has risen over the past couple of years (Kumar and Sharma, 2014). India accounts for biogas production of 2.1 billion m^3/year, which is unprecedentedly low as compared to the appraised range of 29–48 billion m^3/year (Mittal et al., 2018). The possible reason for this failure in achieving the desired range could be the feedstock availability, flaws in policies, use of other renewable energy, lack of subsidies, etc.

The barriers to such technology have a different scenario in urban and rural areas. Small scale biogas plants or family-owned anaerobic digestors are mainly controlled by individual houses. It usually does not provide any kind of monetary benefits; rather, its end products are used as cooking gas by replacing fuelwood and as biofertilizers for enhancing crop yield. Rural India accounts for the highest number of individual biogas plant setup. However, because of expensive technology and limited market, methane upgrading processes and electricity production in rural areas are limited. The large-scale biogas plants are commercialized, which management is in the hand of public-private partnerships or entirely by private. The large-scale plants aim to create a healthy revenue stream by selling produced products for vehicle fuel, electricity or heat and biofertilizers. To increase the efficiency and reduce the cost for biomethanation, constructive steps need to be taken for adequate segregation of food waste and other organic waste at the source. In the rural area, biogas plants are solely dependent on cattle manure and agricultural waste.

The Indian Ministry of New and Renewable Energy (MNRE) promotes several waste-to-energy projects in India. India had launched many supportive schemes like National Biogas and Manure Management Program (NBMMP) to enhance the production of biogas and for the installation of domestic biogas plants in the rural and semiurban regions. In the line of the cleanliness campaign called "Swachh Bharat" or "Clean India" mission, the Government of India has put forward a scheme called "GOBAR-DHAN," which stands for "Galvanizing Organic Bio-Agro Resources" (Vikaspedia, 2020). This scheme aims to empower the farmers of India in making them self-reliant by using agricultural crop residues, including solid and liquid wastes. There are also some other schemes backed by the provincial government, which complements the central scheme in providing subsidies and training for the installation of biogas plants (MNRE, 2020).

The "New National Biogas and Organic Manure Program" (NNBOMP), an improvised and updated version of NBMMP, has been launched recently. It aims to provide clean fuel for various purposes like cooking and meeting the power needs of households in rural and semi-urban areas. The scheme also empowers, engages women, and reduces dependency on forests for fuel, which eventually lowers the emissions of greenhouse gases. Another program called "Biogas-based Power Generation (off-grid) and Thermal Energy Application Program" (BPGTP) aims in promoting power generation (off-grid) based on renewable energy sources with a capacity of 3–250 kW (Mercom India, 2019). The heat generated from biomass-based biogas plants with a size of 30–2500 m^3 can be used for heat exchange applications (Mittal et al., 2018). Many private organizations in India have come forward and started investing in renewable and sustainable energy generation. As of 2017, there were 4.9 million biogas plants installed across India (Statista, 2017).

Several non-governmental organizations have also started using waste for generating energy in India. Akshaya Patra Foundation, one of the globally renowned nongovernmental organizations operating in India, works towards providing a free meal every day to 1.8 million schoolchildren (Akshaya Patra, 2020). The organization is funded by numerous individuals and corporates to operate. The organization has around 52 kitchens for meal preparation and simultaneously generates large volumes of organic wastes. The foundation is committed to the sustainable use of kitchen generated wastes, so they have installed many biogas plants to process it. The biogas produced is then used as fuel for cooking purposes. The Akshaya Patra's kitchen at Ballari, India, has a biogas plant that processes around 1 ton of kitchen wastes per day and produces biogas, which is used as an alternative to liquified petroleum gas.

Apart from cooking fuel, the plant also generates a large volume of organic fertilizer, which eventually could replace the use of harmful chemical fertilizers in farming. Similarly, one of the kitchens at Bangaluru, India produces 1.4 tons of biogas monthly, which is equivalent to 0.7 tons of liquified petroleum gas and saves approximately Indian Rupees 38,500 every month, which is equivalent to the US $520 (as of September 2020). The plant processes approximately 20 metric tons of waste to generate energy (Akshaya Patra, 2020).

Green Elephant India Private Limited, a leading biogas facility in India, also leads in producing compressed biogas. The company's prefabricated biogas setup called the "Green box" has a huge Industrial consumer base in India, the Far East and some countries of Europe. The company is operating its largest plant at Satara, Maharashtra, India, with a waste processing capacity of 1 ton and produces biogas of volume 28 m^3 every day. The plant notably reduces CO_2 emissions (9.7 per annum) (Green Elephant, 2020).

Biogas plants under the MNRE program called the "National Biogas and Manure Management Program" (NBMMP) produce nearly substantial amounts of biogas every day (Mittal et al., 2018). The project employs a thousand people and supports regional-based training and development centers. The Government of India's proactive approach and supportive policies have enabled people and industries to adopt biogas plants on a small and large scale across the country. The cities are yet to develop and understand the numerous opportunities tied with the utilization of waste generated from municipal and industrial sources in biogas and biomethane production. The improved policies and more subsidies with proper mapping can play an immense role in bringing biogas revolution in cities also. If they are provided with appropriate technology, a concrete source for feedstocks, availability of lands and more lucrative subsidies, then developing countries like India will do exemplarily good in the generation of renewable and sustainable energy from waste (MNRE, 2020).

5.6 CONCLUSIONS

Biomethane is one of the significant and economical biofuels that can be produced and used as an alternative for natural gas in heating, electricity, and transportation fuels. With the advent of climate change and overreliance on petroleum resources, biofuels such as biomethane are very useful sources of energy. This book chapter outlines recent advancements in biomethane production. The thermochemical process (i.e., gasification) and biological process (i.e., anaerobic digestion) technologies are the two main routes for producing biomethane from organic materials. While the latter process applies heat and chemical reactions for organic material degradation, the former uses microorganisms and enzymes. Owing to the individual demerits of each process, some researchers have explored an integrated process that incorporates both thermochemical and biological conversion routes.

ACKNOWLEDGMENTS

The authors would like to thank the Natural Sciences and Engineering Research Council of Canada (NSERC) and Canada Research Chairs (CRC) program for funding this bioenergy research.

KEYWORDS

- **anaerobic digestion**
- **biomethane**
- **gasification**
- **thermochemical technologies**

REFERENCES

Akshaya, P., (2020). *Akshaya Patra Adopts First of its Kind Biogas Units*. https://www.akshayapatra.org/news/first-of-its-kind-biogas-units (accessed on 25 June 2021).

Alamia, A., Larsson, A., Breitholtz, C., & Thunman, H., (2017). Performance of large-scale biomass gasifiers in a biorefinery, a state-of-the-art reference. *Int. J. Energy Res., 41*, 2001–2019.

Angelidaki, I., Karakashev, D., Batstone, D. J., Plugge, C. M., & Stams, A. J., (2011). Biomethanation and its potential. In: Rosenzwig, A. C., & Ragsdale, S. W., (eds.), *Methods in Enzymology* (pp. 327–351). Academic Press.

Arrhenius, K., Fischer, A., & Büker, O., (2019). Methods for sampling biogas and biomethane on adsorbent tubes after collection in gasbags. *Appl. Sci., 9*, 1171.

Balussou, D., Kleyböcker, A., McKenna, R., Möst, D., & Fichtner, W., (2011). An economic analysis of three operational codigestion biogas plants in Germany. *Waste Biomass Valor., 3*, 23–41.

Barik, D., Sah, S., & Murugan, S., (2013). Biogas production and storage for fueling internal combustion engines. *Int. J. Emerg. Technol. Adv. Eng., 3*, 193–202.

Bioenergy Consult., (2020). *A Glance at Biogas Storage Systems*. https://www.bioenergyconsult.com/tag/biogas-storage-systems/ (accessed on 25 June 2021).

Budzianowski, W. M., & Brodacka, M., (2017). Biomethane storage: Evaluation of technologies, end uses, business models, and sustainability. *Energy Convers. Manage., 141*, 254–273.

Chen, X., Jiang, J., Tian, S., & Li, K., (2015). Biogas dry reforming for syngas production: Catalytic performance of nickel supported on waste-derived SiO_2. *Catal. Sci., 5*, 860–868.

Fabbri, D., & Torri, C., (2016). Linking pyrolysis and anaerobic digestion (Py-AD) for the conversion of lignocellulosic biomass. *Curr. Opin. Biotech., 38*, 167–173.

Ferella, F., Cucchiella, F., D'Adamo, I., & Gallucci, K., (2019). A techno-economic assessment of biogas upgrading in a developed market. *J. Clean. Prod., 210*, 945–957.

Feroldi, M., Neves, A. C., Bach, V. R., & Alves, H. J., (2016). Adsorption technology for the storage of natural gas and biomethane from biogas. *Int. J. Energy Res., 40*, 1890–1900.

Gallagher, C., & Murphy, J. D., (2013). Is it better to produce biomethane via thermochemical or biological routes? An energy balance perspective. *Biofuel. Bioprod. Biorefin., 7*, 273–281.

Global Market Insights, I, (2020). *World Biogas Market Value to cross $110 Billion by 2025: Global Market Insights, Inc*. https://www.globenewswire.com/news-release/2019/10/03/1924549/0/en/World-Biogas-Market-value-to-cross-110-billion-by-2025-Global-Market-Insights-Inc.html (accessed on 25 June 2021).

Görling, M., Larsson, M., & Alvfors, P., (2013). Bio-methane via fast pyrolysis of biomass. *Appl. Energy., 112*, 440–447.

Hattori, S., (2008). Syntrophic acetate-oxidizing microbes in methanogenic environments. *Microbes Environ., 23*(2), 118–127.

Korres, N. E., Singh, A., Nizami, A. S., & Murphy, J. D., (2010). Is grass biomethane a sustainable transport biofuel?. *Biofuel Bioprod Biorefin., 4*, 310–325.

Krich, K., Augenstein, D., Batmale, J. P., Benemann, J., Rutledge, B., & Salour, D., (2005). *Biomethane from Dairy Waste: A Sourcebook for the Production and Use of Renewable Natural Gas in California*. Western United Dairymen, 2019.

Kumar, A., & Sharma, M. P., (2014). Estimation of GHG emission and energy recovery potential from MSW landfill sites. *Sustain. Energy Technol. Assess., 5*, 50–61.

Li, H., Li, L., Lin, R. B., Zhou, W., Zhang, Z., Xiang, S., & Chen, B., (2019). Porous metal-organic frameworks for gas storage and separation: Status and challenges. *Energy Chem., 1*, 100006.

Li, H., Mehmood, D., Thorin, E., & Yu, Z., (2017). Biomethane production via anaerobic digestion and biomass gasification. *Energy Procedia., 105*, 1172–1177.

Lin, R., Cheng, J., & Murphy, J. D., (2018). Inhibition of thermochemical treatment on biological hydrogen and methane coproduction from algae-derived glucose/glycine. *Energy Convers. Manag., 158*, 201–209.

Lü, F., Hua, Z., Shao, L., & He, P., (2018). Loop bioenergy production and carbon sequestration of polymeric waste by integrating biochemical and thermochemical conversion processes: A conceptual framework and recent advances. *Renew. Energy, 124*, 202–211.

Luz, F. C., Volpe, M., Fiori, L., Manni, A., Cordiner, S., Mulone, V., & Rocco, V., (2018). Spent coffee enhanced biomethane potential via an integrated hydrothermal carbonization-anaerobic digestion process. *Bioresour. Technol., 256*, 102–109.

Matos, I., Bernardo, M., & Fonseca, I., (2017). Porous carbon: A versatile material for catalysis. *Catal. Today, 285*, 194–203.

Mercom India., (2019). *MNRE Seeks Bidders for an Evaluation Study of its Biogas and Thermal Energy Program*. https://mercomindia.com/mnre-seeks-bidders-for-an-evaluation-study-of-its-biogas/ (accessed on 25 June 2021).

Mittal, S., Ahlgren, E. O., & Shukla, P. R., (2018). Barriers to biogas dissemination in India: A review. *Energy Policy, 112*, 361–370.

MNRE, Ministry of New and Renewable Energy-Government of India, (2020). https://mnre.gov.in/ (accessed on 25 June 2021).

Moghaddam, E. A., Ericsson, N., Hansson, P. A., & Nordberg, Å., (2019). Exploring the potential for biomethane production by willow pyrolysis using life cycle assessment methodology. *Energy Sustain. Soc., 9*, 1–18.

Mohanty, P., Nanda, S., Pant, K. K., Naik, S., Kozinski, J. A., & Dalai, A. K., (2013). Evaluation of the physiochemical development of biochars obtained from pyrolysis of wheat straw, timothy grass and pinewood: Effects of heating rate. *Anal. Appl. Pyrolysis, 104*, 485–493.

Nanda, S., Dalai, A. K., Berruti, F., & Kozinski, J. A., (2016a). Biochar as an exceptional bioresource for energy, agronomy, carbon sequestration, activated carbon and specialty materials. *Waste Biomass Valor., 7*, 201–235.

Nanda, S., Kozinski, J. A., & Dalai, A. K., (2016b). Lignocellulosic biomass: A review of conversion technologies and fuel products. *Curr. Biochem. Eng., 3*, 24–36.

Nanda, S., Mohammad, J., Reddy, S. N., Kozinski, J. A., & Dalai, A. K., (2014). Pathways of lignocellulosic biomass conversion to renewable fuels. *Biomass Convers. Bioref., 4*, 157–191.

Nanda, S., Rana, R., Zheng, Y., Kozinski, J. A., & Dalai, A. K., (2017). Insights on pathways for hydrogen generation from ethanol. *Sustain. Energy Fuels, 1*, 1232–1245.

Nanda, S., Reddy, S. N., Mitra, S. K., & Kozinski, J. A., (2016c). The progressive routes for carbon capture and sequestration. *Energy Sci. Eng., 4*, 99–122.

Okolie, J. A., Nanda, S., Dalai, A. K., Berruti, F., & Kozinski, J. A., (2020). A review on subcritical and supercritical water gasification of biogenic, polymeric and petroleum wastes to hydrogen-rich synthesis gas. *Renew. Sust. Energy Rev., 119*, 109546.

Okolie, J. A., Rana, R., Nanda, S., Dalai, A. K., & Kozinski, J. A., (2019). Supercritical water gasification of biomass: A state-of-the-art review of process parameters, reaction mechanisms and catalysis. *Sustain. Energy Fuels, 3*, 578–598.

Pääkkönen, A., Aro, K., Aalto, P., Konttinen, J., & Kojo, M., (2019). The potential of biomethane in replacing fossil fuels in heavy transport—a case study on Finland. *Sustainability, 11*, 4750.

Pore, S. D., Engineer, A., Dagar, S. S., & Dhakephalkar, P. K., (2019). Meta-omics based analyses of microbiome involved in biomethanation of rice straw in a thermophilic anaerobic bioreactor under optimized conditions. *Bioresour. Technol., 279*, 25–33.

Reddy, S. N., Nanda, S., & Kozinski, J. A., (2016). Supercritical water gasification of glycerol and methanol mixtures as model waste residues from biodiesel refinery. *Chem. Eng. Res. Des., 113*, 17–27.

Reddy, S. N., Nanda, S., Dalai, A. K., & Kozinski, J. A., (2014). Supercritical water gasification of biomass for hydrogen production. *Int. J. Hydrogen Energy, 39*, 6912–6926.

Saha, S., Basak, B., Hwang, J. H., Salama, E. S., Chatterjee, P. K., & Jeon, B. H., (2020). Microbial symbiosis: A network towards biomethanation. *Trends in Microbiol.*, 1–17.

Salman, C. A., Schwede, S., Thorin, E., & Yan, J., (2017). Enhancing biomethane production by integrating pyrolysis and anaerobic digestion processes. *Appl. Energy., 204*, 1074–1083.

Statista, (2017). *India: Number of Biogas Plants by State*. https://www.statista.com/statistics/941298/india-number-of-biogas-plants-by-state/ (accessed on 25 June 2021).

Vikaspedia., (2020). *GOBAR-DHAN Scheme*. https://vikaspedia.in/health/sanitation-and-hygiene/gobardhan-scheme (accessed on 25 June 2021).

Wang, S., Bi, X., & Wang, S., (2015). Thermodynamic analysis of biomass gasification for biomethane production. *Energy, 90*, 1207–1218.

Wang, S., Lu, L., Wu, D., & Lu, X., (2017). Biomethane storage in activated carbons: A grand canonical Monte Carlo simulation study. *Mol. Simulat., 43*, 1142–1152.

CHAPTER 6

Thermochemical Methanation Technologies for Biosynthetic Natural Gas Production

KUNWAR PARITOSH,[1] NUPUR KESHARWANI,[2] NIDHI PAREEK,[3] and VIVEKANAND VIVEKANAND[1]

[1]*Center for Energy and Environment, Malaviya National Institute of Technology, Jaipur, Rajasthan, India*
E-mail: vivekanand.cee@mnit.ac.in (Vivekanand Vivekanand)

[2]*Department of Civil Engineering, National Institute of Technology, Raipur, Chhattisgarh, India*

[3]*Department of Microbiology, Central University of Rajasthan, Ajmer, Rajasthan, India*

ABSTRACT

The rigorous use of fossil fuels is leading toward an inevitable situation of environmental degradation. The declining health of the world is of serious concern. However, as an alternative to fossil fuels, biomass or bio-based fuels could be employed as a sustainable development option. Synthetic natural gas or thermochemical methanation of syngas produced from biomass gasification could subside the environmental concerns around the world. Thermochemical methanation of syngas could transform the fossil fuel-based industries and provide a high-density green fuel. However, the impurities present in the syngas after the gasification process could hamper its smooth methanation. The impurities such as particulate matter, tar, sulfur alkalis, and nitrogenous compounds should be removed or reduced to acceptable limits for end-uses. This chapter discusses the gasification process for syngas production and its cleaning for biosynthetic natural gas production. The chapter also sheds light on the use of catalysts for enhanced synthetic natural gas production from syngas.

6.1 INTRODUCTION

The global community is facing challenges to fulfill the energy needs by nonconventional and renewable sources. The exploitation of conventional sources of energy like coal, oil, and gases has changed the environmental aspects of the world. Fossil fuels provide economical energy and power solution along with transportation and chemicals for various applications. However, the ongoing burden on conventional sources due to increased population makes it hard to fulfill energy and power demands on its own. In addition, the excess burden on conventional sources resulted in various emission problems and environmental concerns. As per the International Energy Outlook (2018), major anthropogenic greenhouse gas emissions are due to the excessive use of fossil fuels. Therefore, it is of great concern to end or reduces greenhouse gas emissions while providing energy security simultaneously.

The annual report of the International Energy Agency (IEA), i.e., International Energy Outlook (2017), indicated that between 2015 and 2040, the world energy need could increase by 28%, with half of the increase ascribed to rapidly developing countries like India and China. In addition, fossil fuels are likely to meet the world's 80% energy demand, although renewables and nuclear energy sectors are increasing. Liquid fuels would be sharing most of the quota of fossil fuels as it fulfills the need of the transportation sector as well as have industrial application. However, The IEA predicted that the share of liquid fuel could be decreased by 3% in 2040 as liquid fuel supply largely depends on supply chain management scenarios. Nevertheless, gaseous fuels have been projected as the fastest-growing energy carrier and increasing by 1.4% on an annual basis as compared to liquid fuels, which are growing by 0.7% per year. This report showed that natural gas as gaseous fuel is replacing coal and nuclear for energy generation.

Physical, thermochemical, and biological conversion of lignocellulosic biomass can be employed for energy and fuel production. However, the combustion process of lignocellulosic biomass can lead to the generation of soot particles and polyaromatic hydrocarbons. Pyrolysis, on the other hand, can produce the industrial-grade organic chemicals, bio-oil, and biochar. The thermochemical conversion of lignocellulosic biomass can provide multidirectional end-use products such as syngas. Syngas can be synthesized as methane and used as an energy carrier. In the thermochemical processes (Figure 6.1), the steps included for the conversion of biomass are drying of biomass, impurity removal, cracking, syngas production and cleaning of the syngas.

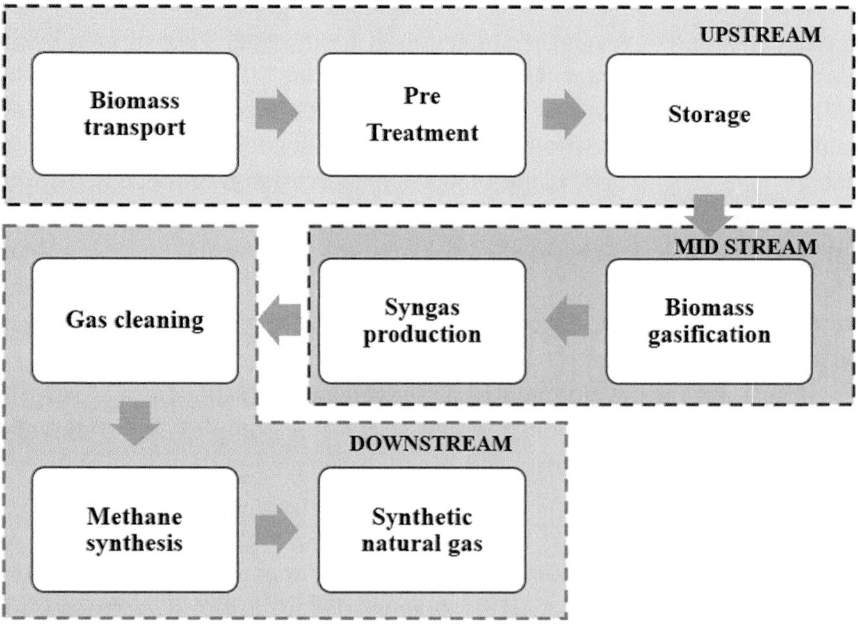

FIGURE 6.1 Synthetic natural gas production by thermochemical conversion of biomass.

6.2 BIOGENIC WASTE AND ENERGY SECTOR

Bio-based feedstock can be thermochemically processed for biosynthetic natural gas (BioSNG) production using a gasifier. The biobased feedstock is generally termed as lignocellulosic biomass, and primarily, it consists of three parts. The cellulose, hemicellulose, and lignin as a defense mechanism against oxidative stress. Energy crops, forestry waste, crop residues, bagasse, and crop processing industries waste are termed as lignocellulosic biomass. The availability and utilization potential could differ from place to place around the world. For example, crop residues are an important source of animal feed in India. In the northern hemisphere of the world, forestry waste possesses huge potential for biomass and biofuels.

The collection of waste feedstock has existed previously in numerous forms. However, as soon as waste has been recognized as a resource for energy application and nutrient recycling, technological advances have been evolved (Iacovidou et al., 2017). Legal, monetary, and environmental aspects play a vital role in the collection of waste for various applications. Typical composition of waste can be depicted as glass, plastic, rubber, and organic matter. However, the collection of food waste is not offered by all local authorities separately (Kumar and Samaddar, 2017). In general, waste collected from different sources is comprised of mixed nature, which could make sorting, and the extra task before the application of waste as a bioresource. The very idea of using biomass resources is to produce non-fossil natural gas and to reduce carbon footprint. Plant-based biomass, a lignocellulosic material, could be used as feedstock in the thermal energy system. However, the potential of biomass as feedstock for thermal energy system varies across the globe. To overcome this, utilization of energy crops such as miscanthus, switchgrass, hemp, and reed canary grass are endorsed.

If sustainable development is practiced extensively, it could help to reduce the risk of greenhouse gases as well as complex environmental issues such as energy security and supply management (Tagliaferri, 2016). For having a carbon emission-free society, infrastructure for the integration of renewable sources for energy carriers and energy utilization as transportation or in the industry is necessary. For achieving the integrated infrastructure, Abanades (2018), suggested decentralized pathways for energy production and use. The carbon emission strongly depends on the technology adopted by the energy sector (Figure 6.2).

Biowaste, a renewable source, could help to decarbonize the energy sector (Bosmans et al., 2013). For several years, sewage sludge was the primary feedstock for gas production. However, energy crops have experimented with the production of biogas. There are primarily two conversion technologies for converting biowaste into energy carriers. These are biochemical and thermochemical conversion techniques. For example, anaerobic digestion and landfill fall into the category of biochemical conversion while gasification and pyrolysis are related to thermochemical conversion technology.

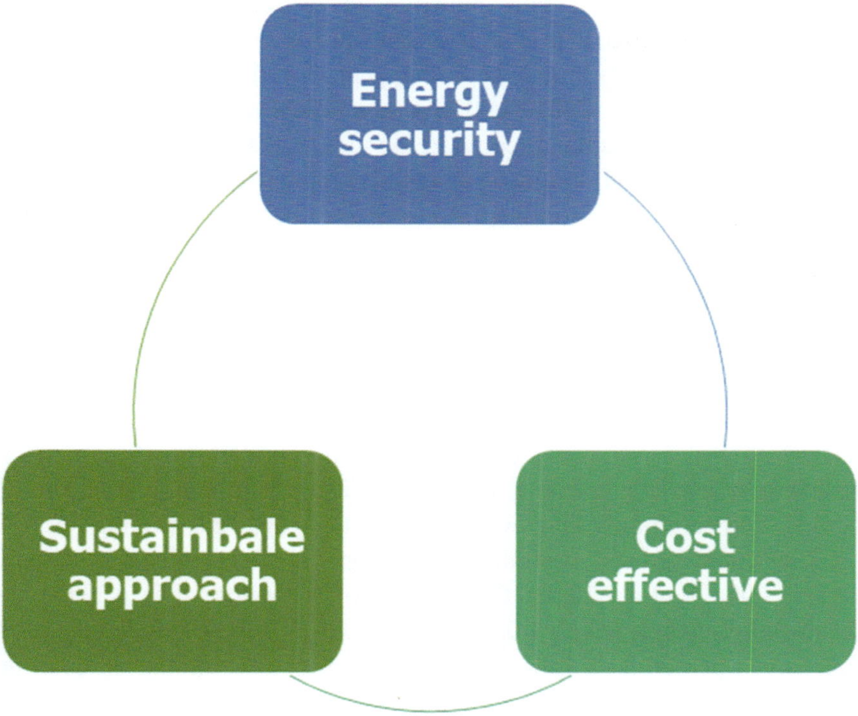

FIGURE 6.2 An interconnected model of energy generation.

TABLE 6.1 Composition of Biogas and Landfill Gas

Constituents	Biogas	Landfill gas
Methane (CH_4)	55–65%	35–65%
Hydrogen (H_2)	Traces	0–2%
Carbon dioxide (CO_2)	35–50%	20–50%
Nitrogen (N_2)	0–1%	5–35%
Ammonia (NH_3)	<100 ppm	<5 ppm
Hydrogen sulfide (H_2S)	<10,000 ppm	<100 ppm

Biochemical conversion technology is a controlled or rather a spontaneous way to convert the waste, which consists of high organic content and a considerable amount of moisture (Table 6.1). Anaerobic digestion, a biochemical conversion technique, could be employed for the multidirectional purpose such as waste stabilization, manure production, production of energy carrier along with industrial waste treatment and utilization (Paritosh et al., 2020). In the anaerobic process, microbes decompose the organic matter and methane is being produced as a by-product. The methane could be used for electricity generation as well as for cooking. Landfill, on the other hand, could be divided into two different manners as open dumping and engineered landfills for gas recovery (Nanda and Berruti, 2021a, 2021b). In a landfill, waste is dumped as layer-by-layer and then covered with suitable material. Furthermore, landfill gas is produced, which could be used for various energy purposes.

6.3 GASIFICATION

Gasification is a process to thermochemically convert the organic material into syngas (Figure 6.3). The gas produced is a combination of CO and H_2 and commonly known as syngas. The syngas gas coming out of a gasifier could be used to produce energy as well as chemicals by several means (Okolie et al., 2019). Gasification of conventional feedstocks has been exercised extensively (Okolie et al., 2020). However, due to the cost-intensive operations of full-scale gasification plants running on coal, no other full-scale plants were planned with the same technology (Watson et al., 2018). The producer gas could be upgraded for hythane (H_2 + CH_4) or syngas (H_2 and CO) (Figure 6.4). The only problem associated with the conversion of producer gas to hythane and syngas is rigorous cleaning steps. Nevertheless, the motive to use these two gases is environmental pressure and urge to phase out conventional fuels. The counterparts for the fossil fuel-based syngas could be produced by employing advanced synthesis processes to produce synthetic natural gas (SNG) or Fischer-Tropsch fuel (Rauch et al., 2015).

Synthetic natural gas or SNG is a renewable option of syngas, which can directly be fed into existing infrastructure. Therefore, the transition from fossil-based gas to bio-based natural gas or SNG could be seen as a huge potential for a sustainable and renewable society. The other benefit of SNG apart from its sustainability aspect is the availability of production technology. In addition, the GobiGas is one of the demo plants showing the production and application of SNG on a broader scale (Thunman et al., 2018).

Biogenic waste requires no other advanced gasifier compared to conventional one and could be fed in the same gasifier. As per the movement of the fluid, three types of gasifiers are commonly known (Table 6.2). These are fixed bed, fluidized bed and entrained flow type of gasifier. The fixed bed gasifiers can be further classified as updraft and downdraft gasifiers (Figures 6.5 and 6.6) as per the direction of flow. In the updraft gasifier system, biomass counters the flow of product gas; hence, it is also termed as a countercurrent gasifier. The features of this type of gasifier are the generation of product gas with high tar content, low exist temperature of gas resulting in higher thermal efficiency and high moisture biomass could be valorized. On the contrary, in a downdraft gasifier, the flow of product gas and biomass are in the same direction; hence, it is also known as a co-current gasifier. Unlike the countercurrent system, the product gas leaves from the bottom of the gasifier.

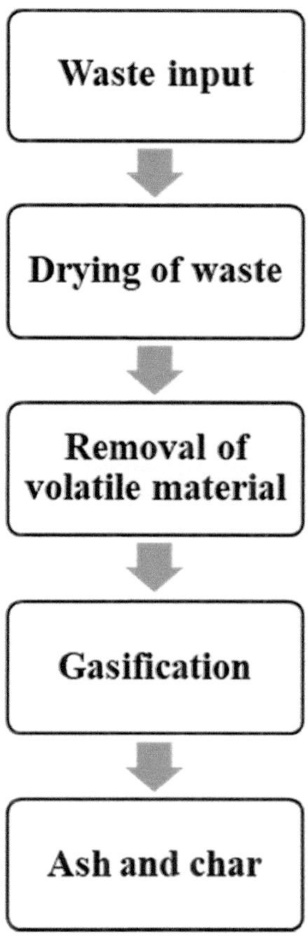

FIGURE 6.3 Schematic of gasification process.

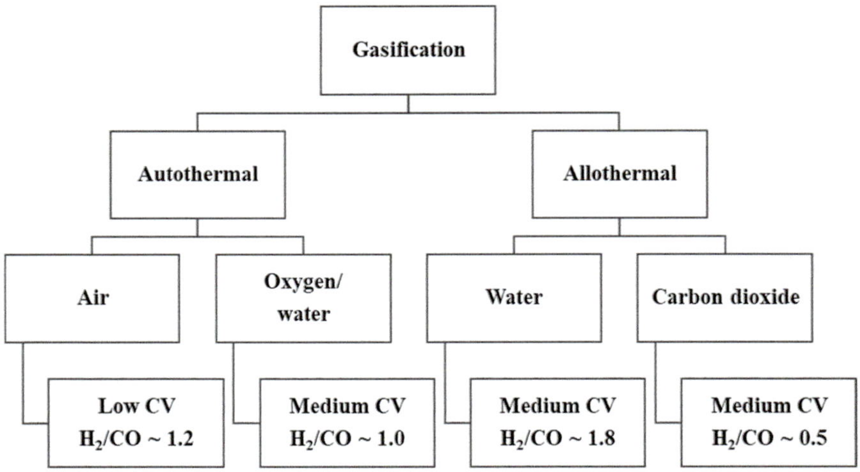

FIGURE 6.4 Effect of gasification agent on calorific value and H_2/CO ratio.

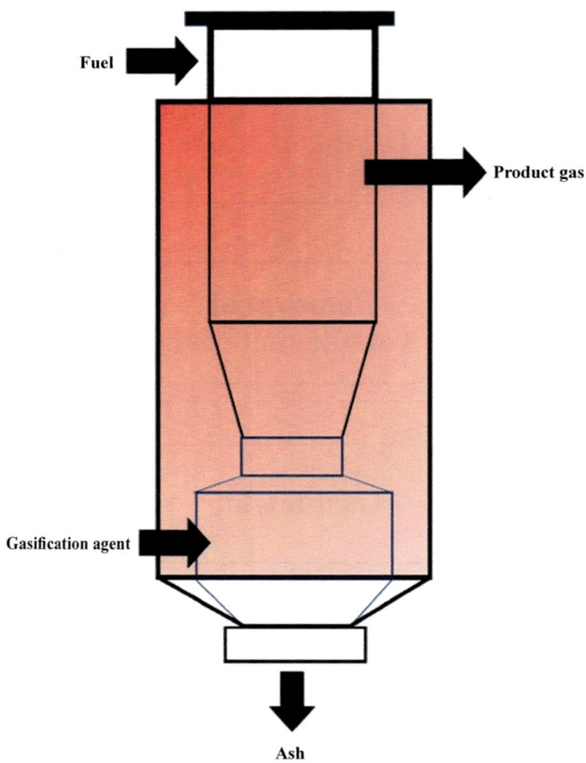

FIGURE 6.5 A fixed bed updraft gasifier unit.

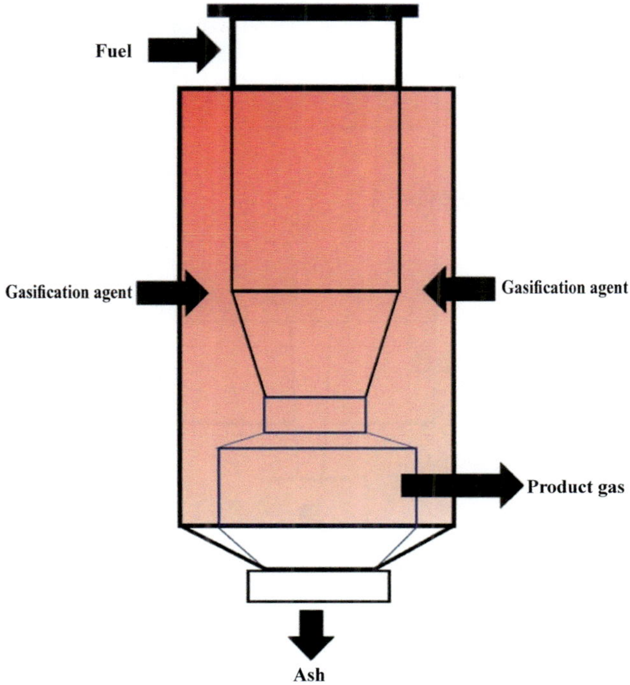

FIGURE 6.6 A fixed bed downdraft gasifier.

The features of the downdraft gasifier are the yield of product gas with low tar content, high exist temperature translating into low thermal efficiency and high moisture containing biomass could not be used. Fluidized bed gasifiers, on the other hand, employ nonconsumable bed material, which can be fluidized by-product gas. The fluidized-bed gasifiers provide mixing and accord better heat and mass transfer as compared to previous ones. In addition, the high heat capacity enabled due to fluidized bed makes it an isothermal system. Furthermore, two types of fluidized bed gasifiers could be divided into bubble type and circulating type fluidized bed gasifier (Figures 6.7 and 6.8). In bubble type fluidized gasifier, the velocity of fluidization is moderate and the bubbles provide effective mixing in the gasifier. In bubble type fluidized bed gasifier, the biomass as fuel could be fed above the bed or directly into the bed. The temperature in this type of gasifier is in the range of 800–900°C and kept almost uniform throughout the gasifier while having a 0.5–1 mm diameter of bed material.

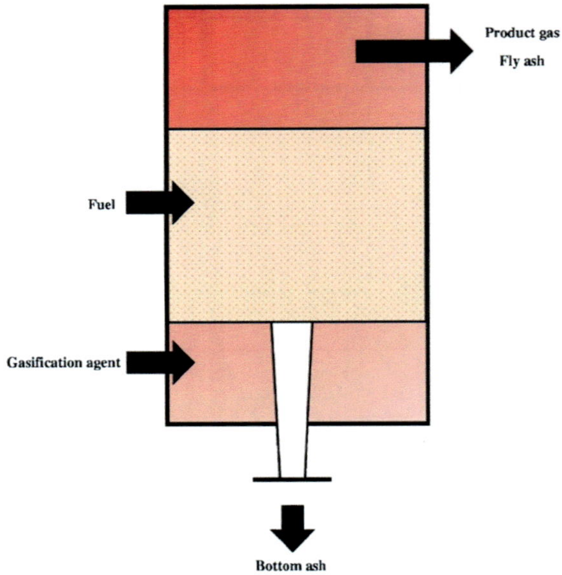

FIGURE 6.7 A bubble type fluidized bed gasifier.

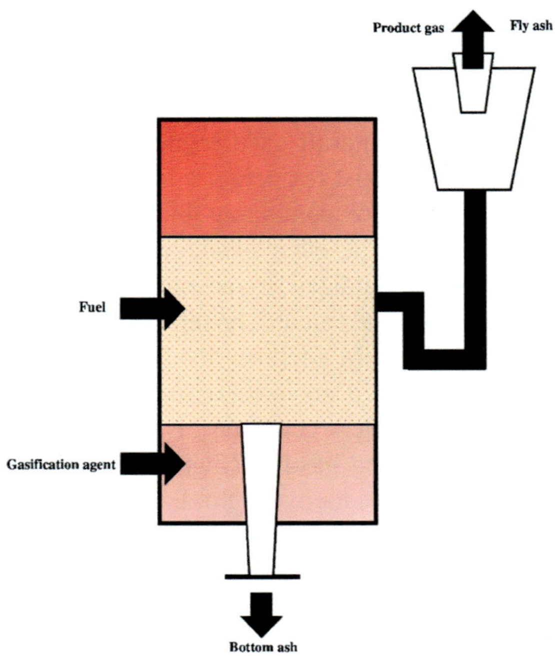

FIGURE 6.8 A circulating type fluidized bed gasifier.

The coarse ash can be extracted from the bed material while fine ash along with the product gas could be collected from the top of the gasifier. The features of bubble-type fluidized bed gasifiers include product gas having medium tar, bed material as a catalyst, high exit temperature translating poor thermal efficiency, able to accommodate heterogeneous material as fuel and both allothermal and autothermal operational mode can be applied. In circulating fluidized bed gasifier, the velocity of fluidization is high. The particles which are leaving the gasifier recycled back into the gasifier after the separation.

TABLE 6.2 Different Gasifiers and Their Pros and Cons

Gasifier	Pros	Cons
Fixed bed gasifier	• High carbon conversion efficiency • High thermal performance • Scale-up is easy • Minimal civil work required	• High amount of tar in syngas • The concentration of CO and H_2 is low in syngas • Catalytic cleaning could be challenging due to poisoning
Fluidized bed gasifier	• Low residence time • Handling of biomass is easy • Better control of temperature • High carbon and thermal conversion efficacy	• Loss of the carbon in ash form • De-fluidization of the bed material could occur at high temperature • High initial cost for set-up
Entrained flow type gasifier	• The temperature profile is uniform throughout the gasifier • Lower tar production at downstream • Easy handling of the gasifier • Scale-up is easy	• A high number of oxidants are required • Pretreatment is required before gasification • High equipment cost • Shorter life span of the system
Rotary kiln gasifier	• Loading is flexible and various biomass could be used • Minimal civil work is required	• Leakage is one of the major issues in rotary kiln gasifier • Lower heat exchange • Maintenance is costly

The power generated per cross-section of the gasifier is considerably high in the circulating type fluidized bed gasifier as compared to the bubble one. The material used in the bed is the same as of bubble-type except for the diameter which is around 0.2–0.5 mm. The features of circulating type fluidized bed gasifier encompass medium tar content with product gas, a low thermal performance due to high exit temperature, possibilities of the high conversion rate of carbon and up to 1 mm particle size fuel can be employed in the gasifier. Other forms of gasifiers are entrained flow gasifiers, rotary drum gasifiers and plasma gasifiers (Figures 6.9–6.11). In an entrained flow gasifier, the thermochemical reaction needed for the gasification process happens in suspension flow (Figure 6.9). The temperature, residence time and pressure for the operation of this type of gasifier are comparatively high (up to 1400°C and 2–5 bar).

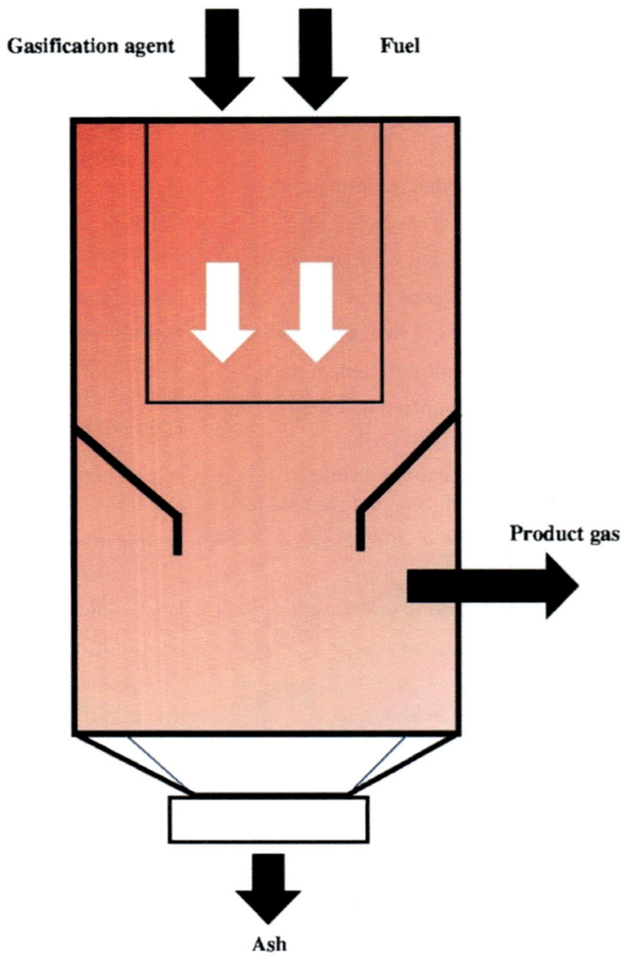

FIGURE 6.9 An entrained flow type gasifier.

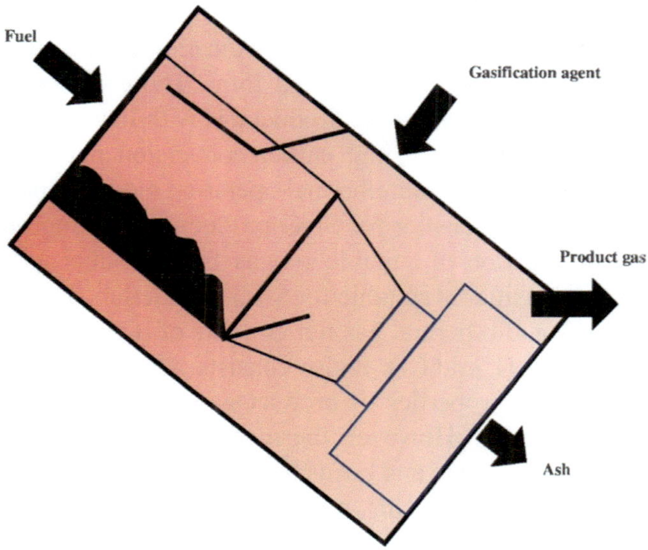

FIGURE 6.10 A rotary drum type gasifier.

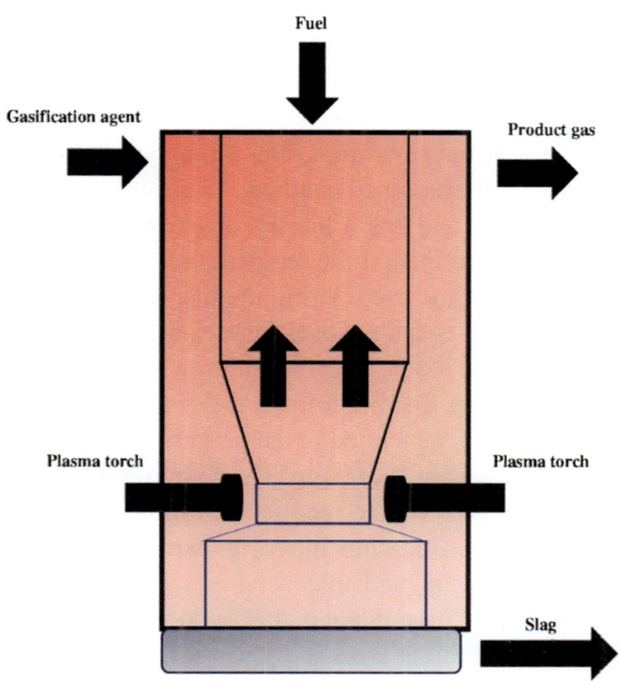

FIGURE 6.11 A single-stage plasma gasifier.

The fuel could be fed in solid form or slurry form from the top long with a gasification agent. Due to high gasification temperature, ash can be collected from the bottom in melted form. The common features of an entrained flow type gasifier are product gas with low methane and tar, medium thermal performance, high carbon conversion and fuel flexibility, vitrified ash generation and autothermal operation mode is applied.

The objective of a gasifier for SNG production running on biogenic waste is the production of suitable syngas from product gas. For this, allothermal type gasifier is suitable for SNG production. The product gas generated from an allothermal gasifier consists of high H_2 content along with a high ratio of H_2 and CO. It also consists of a considerable amount of CH_4 in it. These properties of product gas are very crucial for bio-based SNG production. However, the methane generated is associated with alkanes as well as tars and needs to be removed from the gas stream before application. Besides this, an autothermal gasifier has numerous benefits over allothermal in terms of design and tar content in the gas stream.

6.4 SYNGAS CLEANING FOR SYNTHETIC NATURAL GAS PRODUCTION

The techniques for removal of impurities from syngas could be divided as the cold, wet, and combined method. The cold gas technique is a conventional approach and has a history of effectiveness. The cold gas cleaning generally performed at the temperature lower than the gasification. However, the cost incurred during the disposal of the contaminants to meet the standards is a drawback associated with this technology (Abdoulmoumine et al., 2015). The impurities, which have to be removed, are discussed below. The first impurity in the syngas is particulate matter. Particulate matters (PM) generated in the process of gasification are classically between the range of 1–100 μm. The composition of the particulate matters typically depends on the feedstock of the gasifier (Hoffman and Stein, 2008). The inorganic constituent of particulate matters includes sodium, potassium, calcium and silica.

The classification of particulate matter is done based on aerodynamic diameter. The most common sizes, i.e., 2.5 μm and 10 μm are generally termed as particulate matter $PM_{2.5}$ and PM_{10}. Tar, present in the syngas, consists of condensable organic matters, which could vary from primary compounds to polycyclic hydrocarbons (Figure 6.12). The thermochemical conversion of feedstocks could create different tar species depending on the operating temperature of the gasifier (Devi et al., 2003).

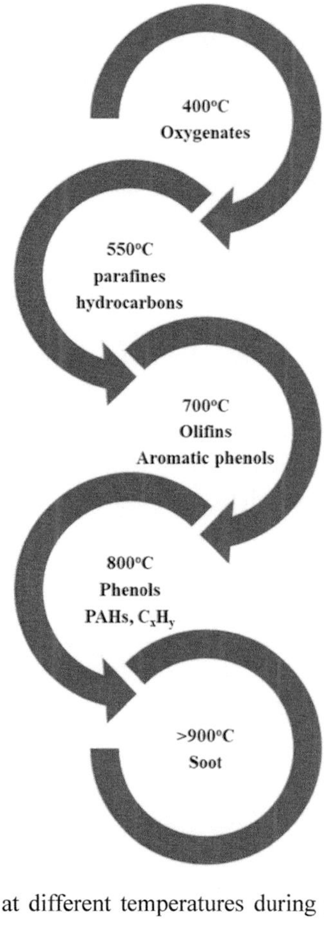

FIGURE 6.12 Tar yield at different temperatures during gasification of lignin-plastic waste.

Bridgwater (1995) reported that the tar content in the case of wood as feedstock is comparatively high to that of coal or peat. In addition, based on the gasifier type, tar content could be low or high. For instance, updraft gasifier roughly yields 15% more tar as compared to downdraft gasifier (Ciferno and Marano, 2002). Characterization of tar is quite difficult as its chemical nature is complex. In a study, it was reported that tar can be classified as a concoction of hydrocarbons having a molecular weight greater than benzene (Maniatis and Beenackers, 2000). Tar can be classified as primary tars released from devolatilizing step in the gasifier and secondary tar released during high temperature and residence time. A further increment in these parameters can result in tertiary tar production.

Sulfur, as a contaminant, mostly present as hydrogen sulfide (H_2S) or sometimes as carbonyl sulfide (COS). As per the feedstock for gasification, the concentration of the H_2S could be between 0.1–30 mL/L of gas. Bio-based feedstock has significantly less concentration of sulfur (0.5 g/kg of biomass) as compared to coal (50 g/kg of coal). However, some biomass such as black liquor waste from the pulp and paper industry could contain more than 1 g of sulfur per kg of biomass (Torres et al., 2007). The concentration of nitrogenous impurity in the syngas heavily depends on feedstock type and process conditions. To avoid oxides of nitrogen as emission from the gas turbine, 0.05 mL/L is required in the syngas stream. Alkaline earth matter and alkali compounds could be the constituent of some feedstock and could release during the gasification process. Alkali compounds form oxides, hydroxide, and sulfates at the temperature above 600°C and could cause corrosion in downstream applications (Turn et al., 1998).

6.4.1 COLD GAS CLEANING TECHNIQUES

In the cold gas cleaning technique, the particulate matter is generally removed by water as a scrubbing agent. The scrubbing techniques used for removing particulate matters are spray, cyclonic spray, impactor, venturi, and wet dynamic scrubbers. However, the basic approach for removal of particulate matter from syngas is using inertial forces to remove it. For removal of particulate matter, in spray scrubber, liquid is dispersed in a counter way or concurrently to the gas flow stream. The removal efficiency of a spray scrubber is around 90% for a particle size of more than 5 μm (Schifftner and Hesketh, 1996). The water plied for the scrubbing of syngas could also remove the gaseous compound that is soluble in water. However, the cost associated with the treatment of wastewater generating form spray scrubber unit makes it a costly approach.

Dynamic scrubbers, on the other hand, have shown the removal efficiency up to 95% for particle size of 5 μm. The mixing of water with the gas stream in the dynamic scrubber is in turbulent mode with the help of mechanical blades. In the impactor scrubber, the syngas passed through the perforated sheets and the sheets are continuously cleaned with the help of water. The removal efficiency of impactor scrubber is around 98% even for large-sized particulate matter (Schifftner and Hesketh, 1996). The static flow model is adopted for the impactor type scrubber and whenever needed. Water circulation is employed to avoid clogging of dirt in the perforated plates. The other particulate matter cleaning technique is the venturi device which works on the principle of reducing the flow area to increasing the gas velocity. However, inertial forces responsible for the removal of particulate matters from syngas could not work on submicron particles.

The tar present in the product gas is detrimental for the instruments and should be removed before the application for SNG production (Materazzi et al., 2014). The problems associated with tar are aerosol formation, condensation in the system and polymerization in the mechanical systems. The proportion of the tar in syngas produced from waste biomass is roughly around 0.1–100 g/Nm^3. However, the proportion of the tar content could be reduced by controlling the operating parameters of a gasifier. For the removal of tars, wet scrubber could work in the same way as for particulate matter. Though the most components of the tar in syngas are not able to get solubilize in water, the wet scrubbing process drops the temperature of the gas stream, and due to this tar present in gaseous form condensed as aerosol on water droplets. Another component of tar is phenol, which could remain in the vapor phase, but it is soluble in the water up to some extent. The wastewater generated from the scrubber unit is recirculated to the processing unit after the removal of the tar from it. However, tar could reduce the effectiveness of water scrubbing.

In the cold gas cleaning technique, conventionally, there are various approaches for sulfur removal from syngas. The solvent method is generally applied for the removal of sulfur. In the chemical-based solvent method, a weak bond between amine and H_2S or CO_2 is created a liquid solvent. Physical absorption of sulfur is also performed for removing it. Methanol and dimethyl ether are a common chemical used for the absorption process. For removal of nitrogenous compounds, water scrubbing is the best option as ammonia is soluble in water. In addition, the condensation of water vapor present in the syngas itself absorbs ammonia downstream. The presence of oxide of sulfur and carbon could enhance the absorption of ammonia. The presence of CO_2 in the syngas improves the formation of aqueous ammonia and thus improves the cleaning process (Bai and Yeh, 1997). Removal of alkali could be performed by lowering the temperature of syngas. Condensation of alkali happens at 300°C and the get mixed in tar or particulate matters in the scrubbing process.

6.4.2 HOT GAS CLEANING

Hot gas cleaning conventionally designed for removal of particulate matters and tars. The hot gas cleaning, with a temperature above 200°C could benefit the thermodynamic processes. In general, the advantages of hot gas cleaning are enhanced syngas conversion efficiency and less waste product in the gas stream. The removal of particulate matter from syngas is based on the principle of inertia, barrier filtration and electrostatics. For the inertial separation of particulates matter from syngas, mass, and acceleration of particles are exploited. For example, a cyclone is the most practical device for inertial separation. Apart from inertial separation, barrier filtration could be used for the removal of particulate matter present in syngas.

In a barrier filtration system, syngas passes through a porous granular solid medium where particulate matter is removed under a different mechanism. These mechanisms are termed as diffusion of the gas, the impact of the gas and settling of the particles under gravitational forces (Seville, 1997). In addition, the porous medium could remove the particles bigger than the size of the material if the particle is going through it. Electrostatic precipitator charges particulate matter present in the syngas stream and particles are separated due to dielectric properties as compared to syngas. An electrostatic precipitator is highly effective in removing the particulate matter from syngas as the electrostatic force is 100 times strong on particles below 30 μm size to that of gravitational force (Lloyd, 1988). The configurations adopted for the electrostatic precipitator are parallel plate and tube type configuration.

In hot gas cleaning, removal of tar is performed as thermal and catalytic cracking, physical separation and nonthermal plasma approach. In a thermal cracking approach for tar removal, small molecule noncondensable gasses are formed from large organic molecules at elevated temperatures. The temperature employed for thermal cracking is in between 1100°C and 1300°C. However, lower temperatures could be effective, as it will provide long residence time to the gas stream. The level of tar in syngas could be reduced by 80% and the reported level of tar by thermal cracking was 15 mg/m^3 at temperatures more than 1200°C (Brandt et al., 2000). Despite being simple and effective, thermal cracking is noneconomical for the downstream process of the gasifier. It was also reported that the application of thermal cracking might increase the concentration of soot particles downstream, which could increase the load on the cleaning process (Houben et al., 2005).

Houben et al. (2002) reported that heating of tar downstream in a fluidized bed reactor polymerized it into soot particles and polycyclic aromatic hydrocarbons. Catalytic cracking requires a temperature range lower than thermal one. In catalytic cracking, the activation energy of the tar is reduced for its decomposition. Due to its property of reducing the temperature for cleaning, the cost associated with it could also be reduced. However, the problems associated with catalytic cracking are the deposition of carbon and fragmentation of tar at downstream (Reddy et al., 2014). The deposition of carbon, also termed as coking, is a phenomenon in which dehydrated carbon particles accumulates on the active site of catalyst and reduce its efficiency (Dayton, 2002).

The geometry of the catalyst could limit the coking process as it changes the surface area of the catalyst. Fragmentation in catalytic cracking, on the other hand, is an outcome of physicochemical forces. Because of the temperature and pressure within the system, the catalyst could be broken into smaller pieces. It is reported that the gas stream might strip off the surface of the metal catalyst downstream (Satterfield, 1997). Besides these, poisoning could also happen while performing catalytic cracking for tar removal. The contaminants in the syngas could be adsorbed on the sites of the catalyst and cause poisoning. Sulfur is one of the contaminants, which cause poisoning in the catalytic cracking process (Bridgwater, 1995).

In hot gas cleaning, H_2S and SO_2 removal is generally focused. Hot gas cleaning mainly focuses on adsorption process using Van der Waals and covalent bonding in physical and chemical adsorption, respectively. These adsorption processes could be reversible and irreversible which could regenerate the sorbent material is reversible. Adsorption of sulfur in the hot gas cleaning process involves three stages. The first stage is called reduction. The adsorbent material is first reduced with sulfur derivative. After this, sulfidation occurs in which metal oxide combines with the sulfur.

At the last stage, the regeneration of sorbent material takes place if it is reversible adsorption. Oxides of metal are best for sulfur adsorption as their chemical properties as best suited for the process. Oxides of zinc, iron, and copper are ranked best for the sulfur adsorption in syngas cleaning (Vamvuka et al., 2004). For ammonia removal in the hot gas cleaning process, decomposition is targeted rather than its removal as in the case of cold gas cleaning. To achieve this, selective thermal decomposition and catalytic oxidation could be adopted for the hot gas cleaning process. Removal of alkali in hot gas cleaning could be achieved by condensation and adsorption. Alkali present in the gas stream could be condensed at a temperature lower than 600°C and removed with particulate matters. On the other hand, the selection criteria of adsorbent material for removing alkali are high-temperature tolerance, high absorptivity and high loading capacity. Hence, kaolinite, diatomaceous earth and clay could be used or could be obtained by synthesis of bauxite as activated alumina.

6.5 SYNGAS METHANATION

Syngas methanation for SNG production is performed as biomass gasification using a well-suited gasifier as per feedstock, removal of impurities, gas cleaning and conditioning. Sabatier and Senderens were the pioneers to discover the reaction for methane synthesis from syngas in the year of 1902. It was proposed that in the presence of nickel as a catalyst, CO, and H_2 reacts to form CH_4 and water molecule provided the temperature is high. The CH_4 synthesis reaction is exothermic. The CH_4 synthesis reaction requires three-part of H_2 and one part of CO. Moreover, for commercial application, CO_2 has to be removed by Rectisol wash (Figures 6.13 and 6.14).

Hydrogen is not sufficient for this reaction and the water-gas shift is employed for this purpose. The catalyst used for methane production usually belongs to group 8 of the periodic table. The activity of iron is more to that of nickel, but economy favors nickel (Vannice 1979). Further, for improving the performance of nickel as a catalyst, doping of nickel performed using molybdenum, tungsten or cobalt. The features of the catalyst to be kept in the mind are its activity, selectivity, and deactivation property. Catalyst deactivation in methane synthesis process from syngas is the loss of catalytic ability. The decrease in the catalytic activity is inevitable over a long period. However, a rapid decline in the ability is a matter of great concern. The possible reason for a catalyst to be deactivated rapidly are impure feedstock in the gasifier and poor process of gasification. A catalyst in the methane synthesis process could be deactivated by several means such as deposition of carbon or ash on the active site, sintering at high temperature, vapor formation in the downstream and inactive phase formation by reacting with the solids present as an impurity.

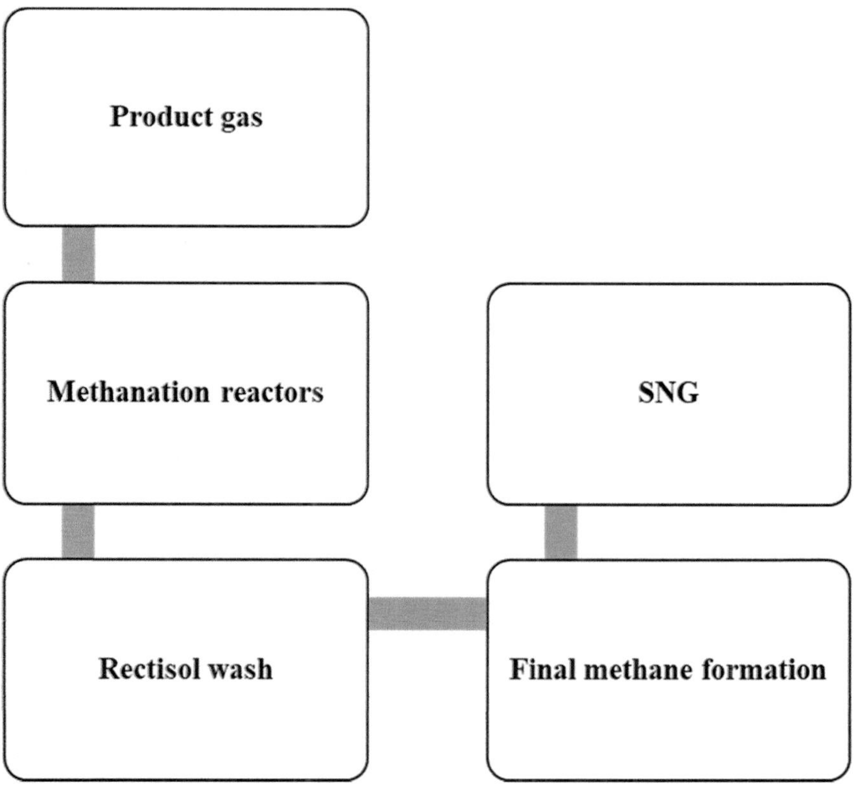

FIGURE 6.13 Single-stage fixed bed methanation flow diagram.

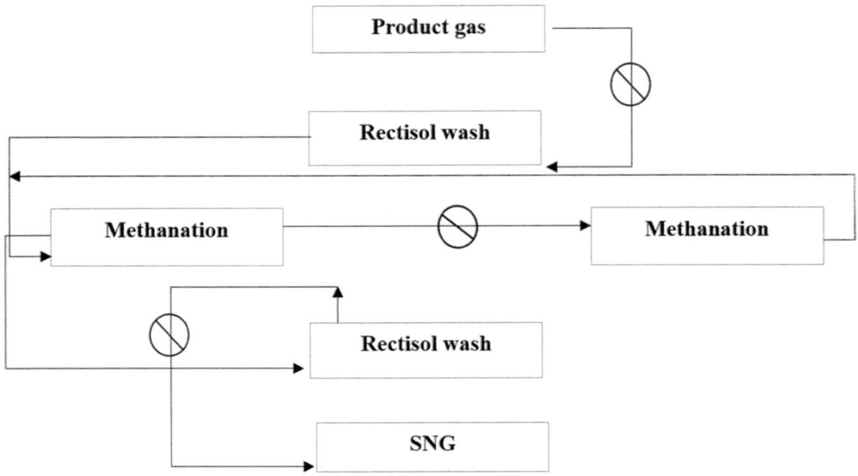

FIGURE 6.14 Multi-stage fixed bed methanation flow diagram.

The deposition of carbon could occur on the catalyst if the catalytic transformation of any hydrocarbon is taking place. In carbon deposition, active sites of a catalyst blocked by the carbon and ash present in the gas. Sometimes, polymeric carbon present in the syngas could encapsulate the active site of the catalyst. In addition, in the catalytic deactivation by carbon deposition, whisker formation could happen to rupture the structure of the catalyst. The sources of the carbon deposition majorly include coke produced by the dissociation of CO and decomposition of hydrocarbons present on syngas. The CO could decompose as CO_2 and carbon exothermically if there is a lack of H_2 in the stream. Apart from this, the formation of coke propagates in different ways. Aliphatic, polyaromatic, aromatic, and olefinic hydrocarbons favor the formation of coke. The hydrocarbon present on the gas stream get absorbed on the surface of catalyst split out H_2 over it. After this, polymerization of coke easily happens on the catalyst.

Sintering is the denaturing of the active site of a nickel catalyst. The thermal shock shrinks the active site of nickel, and this is a temperature-dependent phenomenon. The sintering process is low at a lower temperature range, while at high temperatures, it accelerates. For determining the degree of sintering, X-ray diffraction or scanning electron microscopy is performed as a direct determination technique. For indirect determination of sintering, H_2 chemisorption could be employed. On the other hand, sulfur has a strong affinity towards nickel catalysts in poisoning it by forming strong bonds.

The poisoning of catalysts with sulfur happens in two steps. In the first step, sulfur blocks up to four atoms of nickel by chemical bonding, and in the second step, by a previously established chemical bond, it affects the neighboring nickel atoms. Due to this phenomenon, the atoms of nickel-based catalysts lose their ability to work. Furthermore, the nature of the deactivation of nickel-based catalysts due to sulfur bonding is irreversible.

6.6 CONCLUSIONS

Biomass could decarbonize the energy sector if used tactically. The abundantly available lignocellulosic biomass can curb greenhouse gas emission significantly and could substitute the fossil-based gasifier for syngas production. Further, the syngas could be synthesized to produce SNG. SNG produced from the gasifiers could provide the alternative of fossil-based fuels. However, the selection of the best-suited gasifier is crucial for the lignocellulosic biomass. Some type of gasifier could produce the syngas with heavy tar content, while some could be noneconomical. In addition, the cleaning process of the syngas could be uneconomical in the long run. Further research should be focused on low tar technologies from biomass gasification and onsite methane synthesis for SNG production.

KEYWORDS

- **biogenic waste**
- **biosynthetic natural gas**
- **carbonyl sulfide**
- **energy security**
- **gasification**
- **hydrogen sulfide**
- **International Energy Agency**
- **methane**
- **particulate matters**
- **syngas cleaning**
- **synthetic natural gas**

REFERENCES

Abanades, A., (2018). Natural gas decarbonization as tool for greenhouse gases emission control. *Front. Energy Res., 6*, 47.

Abdoulmoumine, N., Adhikari, S., Kulkarni, A., & Chattanathan, S., (2015). A review on biomass gasification syngas clean up. *Appl. Energy., 155*, 294–307.

Bai, H., & Yeh, A. C., (1997). Removal of CO_2 greenhouse gas by ammonia scrubbing. *Ind. Eng. Chem. Res., 36*, 2490–2493.

Bosmans, A., Vanderreydt, I., Geysen, D., & Helsen, L., (2013). The crucial role of waste-to-energy technologies in enhanced landfill mining: A technology review. *J. Clean. Prod., 55*, 10–23.

Brandt, P., Larsen, E., & Henriksen, U., (2000). High tar reduction in a two-stage gasifier. *Energy Fuel, 144*, 816–819.

Bridgwater, A. V., (1995). The technical and economic feasibility of biomass gasification for power-generation. *Fuel, 74*, 631–653.

Ciferno, J. P., & Marano, J. J., (2002). *Benchmarking Biomass Gasification Technologies for Fuels, Chemicals and Hydrogen Production.* U.S. Department of Energy. National Energy Technology Laboratory.

Dayton, D., (2002). *Review of the Literature on Catalytic Biomass Tar Destruction* (p. 33). National Renewable Energy Laboratory.

Devi, L., & Ptasinski, K. J. J., (2003). A review of the primary measures for tar elimination in biomass gasification processes. *Biomass Bioenergy, 24*, 125–140.

Hoffmann, A. C., & Stein, L. E., (2008). *Gas Cyclones and Swirl Tubes: Principles, Design, and Operation* (2nd edn.). Springer, Berlin: New York.

Houben, M. P., De Lange, H. C., & Steenhoven, A. A., (2005). Tar reduction through partial combustion of fuel gas. *Fuel, 84* 817–824.

Houben, M. P., Verschuur, H., Neeft, J., & Ouwens, C., (2002). An analysis and experimental investigation of the cracking and polymerization of tar. *12th European Conference on Biomass for Energy, Industry and Climate Protection*. Netherlands: Amsterdam.

Iacovidou, E., Velis, C. A., Purnell, P., Zwirner, O., Brown, A., & Hahladakis, J., (2017). Metrics for optimizing the multidimensional value of resources recovered from waste in a circular economy: A critical review. *J. Clean. Prod., 166*, 910–938.

International Energy Outlook (IEO), (2018). U.S. Energy Information Administration. Available from: https://www.eia.gov/ (accessed on 25 June 2021).

Kumar, A., & Samadder, S. R., (2017). A review on technological options of waste to energy for effective management of municipal solid waste. *Waste Manag., 69*, 407–422.

Lloyd, D. A., (1988). *Electrostatic Precipitator Handbook* (Vol. XIV, p. 239). Bristol; Philadelphia: A. Hilger.

Maniatis, K., & Beenackers, A. A. C. M., (2000). Tar protocols. IEA bioenergy gasification task. *Biomass Bioenergy, 18*, 1–4.

Materazzi, M., Lettieri, P., Mazzei, L., Taylor, R., & Chapman, C., (2014). Tar evolution in a two stage fluid bed-plasma gasification process for waste valorization. *Fuel Process. Technol., 128*, 146–157.

Nanda, S., & Berruti, F., (2021a). A technical review of bioenergy and resource recovery from municipal solid waste. *J. Hazard. Mater., 403*, 123970.

Nanda, S., & Berruti, F., (2021b). Municipal solid waste management and landfilling technologies: A review. *Environ. Chem. Lett.*, 19, 1433–1456.

Okolie, J. A., Nanda, S., Dalai, A. K., Berruti, F., & Kozinski, J. A., (2020). A review on subcritical and supercritical water gasification of biogenic, polymeric and petroleum wastes to hydrogen-rich synthesis gas. *Renew. Sust. Energy Rev., 119*, 109546.

Okolie, J. A., Rana, R., Nanda, S., Dalai, A. K., & Kozinski, J. A., (2019). Supercritical water gasification of biomass: A state-of-the-art review of process parameters, reaction mechanisms and catalysis. *Sustain. Energy Fuels, 3*, 578–598.

Paritosh, K., Balan, V., Vijay, V. K., & Vivekanand, V., (2020). Simultaneous alkaline treatment of pearl millet straw for enhanced solid-state anaerobic digestion: Experimental investigation and energy analysis. *J. Clean. Prod., 252*, 119798.

Rauch, R., Hrbek, J., & Hofbauer, H., (2015). Biomass gasification for synthesis gas production and applications of the syngas. In: Lund, P. D., Byrne, J., Berndes, G., & Vasalos, L. A., (eds.), *Advances in Bioenergy: The Sustainability Challenge.*

Reddy, S. N., Ding, N., Nanda, S., Dalai, A. K., & Kozinski, J. A., (2014). Supercritical water gasification of biomass in diamond anvil cells and fluidized beds. *Biofuels, Bioprod. Bioref., 8*, 728–737.

Satterfield, C. N., (1996). *Heterogeneous Catalysis in Industrial Practice* (2nd edn., Vol. XVI, p. 554). Malabar, Fla: Krieger Pub.

Schifftner, K. C., & Hesketh, H. E., (1996). *Wet Scrubbers* (2nd edn., p. 206). Lancaster, Pa: Technomic.

Seville, J. P. K., (1997). *Gas Cleaning in Demanding Applications* (1st edn., p. 308). London; New York: Blackie Academic & Professional.

Tagliaferri, C., (2016). *Life Cycle Assessment of Shale Gas, LNG and Waste in the Future UK Energy Mix*. University College London.

Thunman, H., Seemann, M., Vilches, T. B., Maric, J., Pallares, D., Ström, H., Berndes, G., et al., (2018). Advanced biofuel production via gasification lessons learned from 200 man-years of research activity with Chalmers' research gasifier and the GoBiGas demonstration plant. *Energy Sci. Eng., 6*, 6–34.

Torres, W., Pansare, S. S., & Goodwin, J. G., (2007). Hot gas removal of tars, ammonia, and hydrogen sulfide from Biomass gasification gas. *Catal. Rev. Sci. Eng., 49*, 407–456.

Turn, S. Q., Kinoshita, C. M., Ishimura, D. M., & Zhou, J., (1998). The fate of inorganic constituents of biomass in fluidized bed gasification. *Fuel, 77*, 135–146.

Vamvuka, D., Arvanitidis, C., & Zachariadis, D., (2004). Flue gas desulfurization at high temperatures: A review. *Environ. Eng. Sci., 21*, 525–547.

Vannice, M. A., & Garten, R. L., (1979). Metal-support effects on the activity and selectivity of Ni catalysts in CO/H_2 synthesis reactions. *J. Catal., 56*, 236–248.

Watson, J., Zhang, Y., Si, B., Chen, W. T., & De Souza, R., (2018). Gasification of biowaste: A critical review and outlooks. *Renew. Sust. Energy Rev., 83*, 1–17.

CHAPTER 7

A Brief Overview of Fermentative Biohythane Production

PRAKASH K. SARANGI[1] and SONIL NANDA[2]

[1]Directorate of Research, Central Agricultural University, Imphal, Manipur, India
E-mail: sarangi77@yahoo.co.in (Prakash K. Sarangi)

[2]Department of Chemical and Biological Engineering, University of Saskatchewan, Saskatoon, Saskatchewan, Canada

ABSTRACT

Overdependence on fossil sources for fuels and energy creates massive environmental problems that need a paradigm shift for the exploration of renewable biomass resources for solving the energy crisis. The utilization of microbial resources along with waste biomass is a promising source to produce future biofuels like hydrogen and biomethane. Various production methods are adapted for the production of biohydrogen and biomethane. The two-stage fermentation method of waste biomass yields both biomethane and biohydrogen releasing biohythane. A brief description of the production and potential of biohythane are summarized in this chapter.

7.1 INTRODUCTION

Extensive uses of fossil-based fuels in automobiles and industrial sectors, which create great environmental pollution, are the major concern around the globe as far as the energy point is concerned. There are many dependencies on fossil fuels for energy and chemicals (Ghimire et al., 2015).

Such nonrenewable sources are in a diminishing stage in the coming future. In this context, exploration of renewable and green energy sources is the need of the world. Biobased resources are promising options for fuel sources towards sustainability. Waste biomass has a great potential to suffice global energy requirements. Residues of the forest, agriculture, and other biobased sources are harnessed to produce biofuels. Also, such sustainable-based sources aid in economic development with an eco-friendly environment (Sarangi et al., 2020b). Thus, exploration of novel renewable sources can be emphasized to produce energy and biochemicals for a long period in a sustainable manner.

Waste biomass is constituted by a variety of crops, crop residues, and biomass generated from the forest. Many biomasses have been intensively researched for their prospective to be as a raw substance to produce second-generation bioethanol. Potential sources among them are wheat straw, rice husk, rice straw, peels of fruits, sugarcane bagasse, corn cobs, shells of coconut and groundnut, etc. (Vaid and Bajaj 2017). Around the world, renewable, green bioresources have been explored to produce different biofuels such as bioethanol, biobutanol, biopropanol, biogas, biodiesel towards mitigation of fossil-based fuels (Nanda et al., 2020; Pasin et al., 2020; Bhatia et al., 2020; Sarangi et al., 2020). Production of various biofuels has been attained by harnessing waste biomass by a wide range of microorganisms. Such bioenergy sources are bioethanol (Sarangi and Nanda 2020b), biobutanol (Sarangi and Nanda 2018), biomethanol (Sarangi et al., 2020) and biohydrogen (Sarangi and Nanda 2020a) which have long applications in industrial and automobile sectors.

Due to the presence of celluloses and hemicelluloses in waste biomass, it has been focused on the production of biohydrogen and other various useful products (Kumar et al., 2017; Yan et al., 2017). Pretreatment of biomass can release lignin part from such biomass. Hence, bioconversion is the potential method for hydrogen generation for the sustainable supply of H_2 with less environmental pollution with high efficiency (Wu and Chang 2007). Hydrogen has gained wide applications in the fuel and energy sectors. Due to the harmful impacts of fossil fuels, hydrogen generation is largely emphasized for mitigation of the energy crisis. Hydrogen can be termed as the fuel of the future due to many advantages with more net calorific value than other important fuels of fossil origin (Grimes et al., 2008). Hydrogen has about 2.75 times greater energy content than conventional hydrocarbon fuels (Faloye et al., 2014, Ghimire et al., 2015).

The fermentative production method is the most attractive and potential method for hydrogen production (Lin et al., 2012) having feasibility in the hydrogen economy.

A wide variety of organic substances along with wastewaters and algal biomass categories are the basic feedstocks for biohydrogen production processes. Microbial degradation of organic waste without oxygen is known as anaerobic digestion. The conversion of organic substances to CO_2 and CH_4 occurs beside a series of biochemical reactions during an anaerobic process (Bailey and Ollis, 1986). The outcome of such a reaction initiates the breakdown of organic matters during the process of digestion, and the only possibility is because of anaerobic microorganisms. Production of methane by anaerobic digestion could be an economic advantage over aerobic digestion as methane gas can be used to run an effluent treatment plant.

"Hythane" is a term used for both hydrogen and methane as a mixture. Though both gases are produced from nonrenewable sources like natural gas and petroleum. But the term biohythane can be used for the production of hythane, particularly through fermentative ways from organic feedstuffs (Cavinato et al., 2010). Co-production of bio-H_2 and bio-CH_4 through two-stage anaerobic digestion has gained a lot of potential bringing 8–43% enhancement in the recovery of energy. Based on their harmonizing characteristics, both production of hydrogen and methane in the form of biohythane as a mixture is focusing more emphasis than individual production. Biohythane is regarded as an improved transportation fuel as compared to compressed natural gas (CNG) with many characteristics like high flammability range, reduction in ignition temperature and absence of nitrous oxide, etc. On the other hand, biohydrogen and biomethane could be united to produce biohythane that has boosted efficiency towards the combustion process (Porpatham et al., 2007). In this chapter, a detailed description of the production of biohythane from various feedstocks is summarized. Also, the process technologies about biohythane are briefly described.

7.2 BIOHYTHANE

Among all biofuels, biohydrogen, and biohythane have the promising potential for future energy supply sustainably due to enhanced conversion efficiency with low pollutants generation (Hallenbeck and Ghosh, 2009). As far as biohydrogen production, dark fermentation displays a high hydrogen production rate having practical utility in bioenergy scenarios (Angenent et al., 2004). Additionally, biohydrogen through dark fermentation shows the quick growth rate of bacteria, high hydrogen production potential, no oxygen limitation issues along low principal investment (Hawkes et al., 2002; Levin et al., 2004). Wide varieties of waste biomass like starchy biomass along with other wastewaters can be explored to produce biohydrogen. Nevertheless, dark fermentation has a limitation of low substrate conversion efficacy. In this case, about the energy level of 7.5–15% stored in theses, biomass is transformed to biohydrogen and the remaining energy is available as H_2 effluent in the form of volatile fatty acids (VFA) such as butyric acid and acetic acid along with alcohols (Hallenbeck and Ghosh, 2009).

Biohythane has gained promising attraction worldwide as far as the applications as a vehicle fuel are concerned. It stands as the alternative fuel from the degradation of waste biomass and a substitute for nonrenewable hythane (Kongjan et al., 2011). Generally, hythane gas was obtained from natural gas as a base material through a thermochemical process with high-energy use depending on fossil fuel. Hence, the limitations of dark fermentation should be address for economic feasibility. In this context, the degradation of VFA, lactate, and bioalcohols into methane via anaerobic digestion must be accomplished so that the energy efficiency of biomass can be fulfilled completely. Additionally, the combination of both dark fermentation and anaerobic digestion should be attained for efficient conversion of biomass towards the generation of hydrogen and methane as a mixture known as biohythane, through two-stage anaerobic fermentation (Luo et al., 2011; Kongjan et al., 2011).

7.3 TWO-STAGE FERMENTATION PROCESS

Production of both biohydrogen and biomethane from waste biomass is accomplished by dark fermentation and anaerobic digestion, respectively. By combining these two processes can produce both H_2 and CH_4 as a mixture called hythane with 10–15% H_2, 50–55% CH_4 and 30–40% CO_2 (Mamimin et al., 2015) that can be refined to renewable-based hythane by removal of CO_2. The produced lactic acid, VFA, and alcohols can be converted into CH_4 and CO_2 during the second stage.

Dark fermentation is the most potential process for hydrogen generation (Kumar et al., 2017; Łukajtis et al., 2018) by various microorganisms from waste biomass. A mixture of gases along with H_2 and CO_2 are produced (Kotsopoulos et al., 2006; Temudo et al., 2007; Datar et al., 2004; Najafpour et al., 2004). Various microorganisms such as *Enterobacter* spp., *Bacillus* spp., and *Clostridium* spp., have the potential to produce hydrogen from cellulosic substances (Levin et al., 2004). Bacteria can convert glucose to pyruvic acid, which is later converted to CO_2 and H_2 (Figure 7.1) through this method. The first phase has a pH value of 5–6 with a hydraulic retention time of 1–3 days, which are appropriate for acetogens towards the degradation of wastes to H_2.

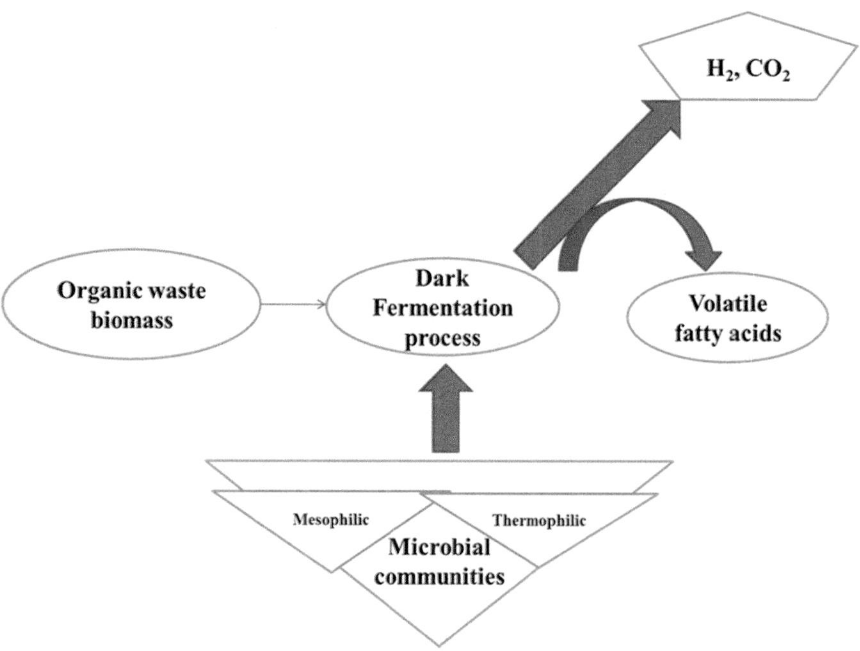

FIGURE 7.1 Production of biohydrogen through dark fermentation.

On the other hand, through the second phage, the produced acetic acid of the first phase can be transformed to CH_4 and CO_2 by the action of acetoclastic methanogens (Figure 7.2) (Kongjan et al., 2011). Also, remaining volatile acids and alcohols are transformed by acetogens to H_2 and CO_2 that can be converted to biomethane by the action of hydrogenotrophic methanogens (Kongjan et al., 2011). Overall, two phasic anaerobic fermentation processes for the generation of biohythane is shown in Figure 7.3.

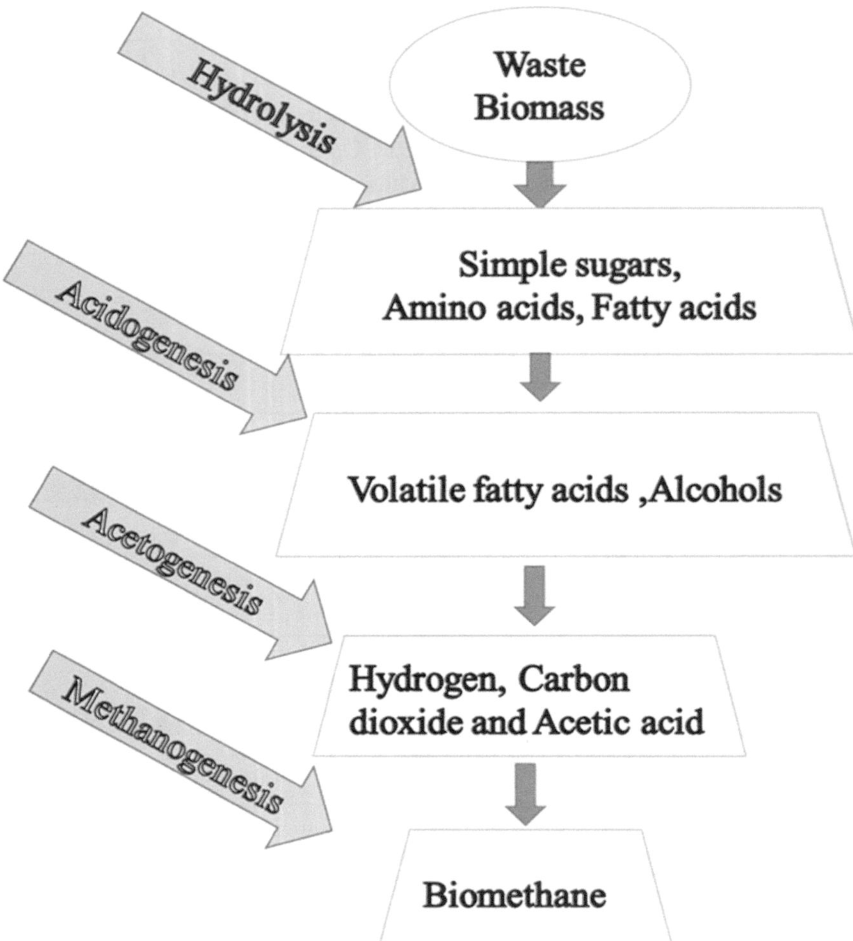

FIGURE 7.2 Production of biomethane from waste biomass.

Through two-stage anaerobic methods, some parameters are benefited such as reactor stability, enhanced energy retrieval, purity of gas products, conversion efficiency and rate of methane production as compared with separate fermentation method (O-Thong et al., 2016). Additionally, major advantages of biohythane have been detected as enhanced energy recapture and reduced fermentation time having more environmentally benign over traditional methods (Liu et al., 2013; Si et al., 2016). Therefore, the integration of biohydrogen with biomethane process worth has great commercial applications with clean fuel for vehicles (Liu et al., 2018).

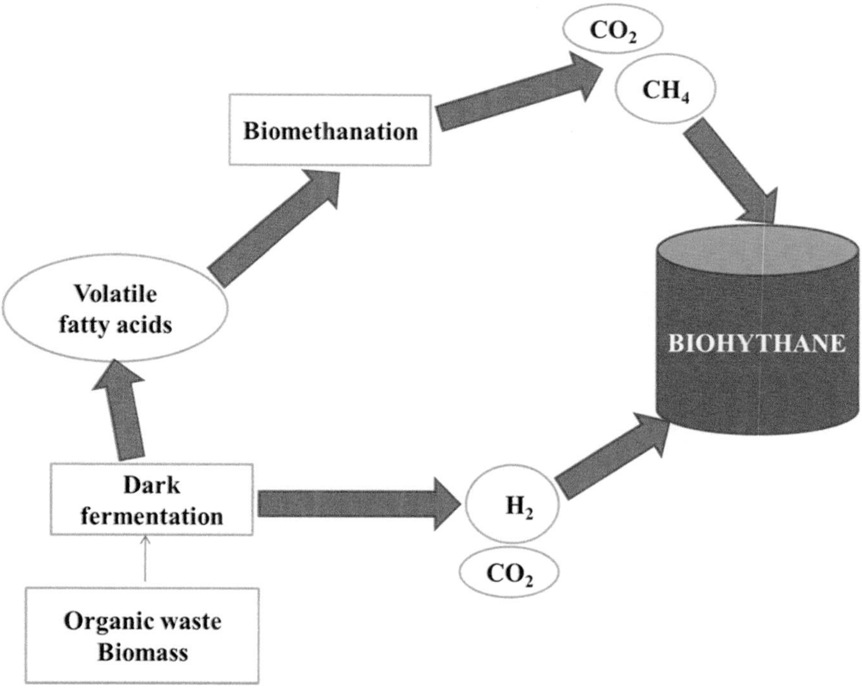

FIGURE 7.3 Production of biohythane through anaerobic fermentation.

The period for the overall production of biohythane production is about 13–18 days (Liu et al., 2013) which depends on the nature of microorganisms. The first stage of fermentation occurs with the help of acidogenic bacteria like *Caldicellulosiruptor* sp., *Enterobacter* sp., *Clostridium* sp., *Thermotoga* sp., and *Thermoanaero* sp. (Nandi and Sengupta, 1998). On the other hand, the second group of bacteria such as methanogenic bacteria like *Methanosarcina* and *Methanoculleus* helps the final production of biohythane (Sompong et al., 2016).

Biohythane production includes two distinct processes in which the first stage has hydrolysis as well as acidogenesis. Through the hydrolysis process, bioconversion of the complex carbohydrates, proteins along with lipids occur. This method leads to the generation of simple sugars, amino acids and long-chain VFAs by fermentative bacteria by catalytic actions. The final products through the first stage of fermentation include H_2, CO_2, VFAs, and lactic acid along with alcohols. Hence, degradation of produced nongaseous products from the first stage can take place through the second stage that involves the production of CH_4 as well as CO_2 (Liu et al., 2008). The overall biohythane production is the function of the metabolites of biohydrogen production path as they can be determined by loading, conversion path, and uniformity in process of the second stage (Wang and Zhao 2009). The real conversion rate of VFA into acetic acid can impact the quantity of microorganisms through the methanogenic path that finally determines the rate of conversion acetate to produce biomethane also. The pH value of about 5–6 with a hydraulic retention time of 1–3 days leads to more hydrogen generation (Liu et al., 2018). During the later stage, acetoclastic methanogens form biomethane and CO_2 from acetic acid by anaerobic processes at pH about a range of pH 7 to 8 with the hydraulic retention time of about 10–15 days (Lee et al., 2010). Wide varieties of biomass such as cellulose-based biomasses, along with food wastes are very suitable for biohythane production (Mamimin et al., 2017). By using a continuously stirred tank reactor digester, biohythane is shown to be in the highest amount of hydrogen (3.2 mol) and methane (3.6 mol) (Sompong et al., 2018). Biohythane production is projected for the actual energy retrieval of 67.70% by conversion of waste biomass (Si et al., 2016).

7.4 CONCLUSIONS

Biohythane generation by microbes is a global concern as sustainability point is concerned with using waste biomass. Microbial biomass should be expanded in their research for its production. Research about biohythane is found to be in the pioneer stage of research. Presently, the research about biohythane has been focused on the lab scale that finds some merits and feasibility along with variable environmental and experimental parameters. Whilst such global research has put some interesting findings and facts which can lead more exploring conclusions and discoveries for large scale. Having a wide range of applications based on a wider range of feedstocks, such research will explore towards mitigation of the energy crisis. Hence, there is necessities more focused research around the world with required optimization in process design along with the operational process of biohythane generation.

Although at a pioneering stage, the production level of biohythane can be explored during basic and advanced research standards having a green engineering approach that can address the bottlenecks of real applications. In addition, economic aspects should be focused on complete process engineering aspects. Nevertheless, energy stabilities should be sought for the extremely positive effect that can benefit on costs appraising large-scale production. Therefore, designs of biohythane production systems have a great role that can put a positive energy-improvement with energy inputs. Additionally, some biotechnological tools are to be applied presently for maximum biohythane production. Microbial culturing conditions as well nature and concentration of feedstock may be focused for maximum recovery.

KEYWORDS

- **biohydrogen**
- **biohythane**
- **biomethane**
- **break-even point**
- **dimethyl ether**
- **discounted cash flow**
- **Fischer-Tropsch process**
- **waste biomass**

REFERENCES

Angenent, L. T., Karim, K., Al-Dahhan, M. H., Wrenn, B. A., & Espinosa, R. D., (2004). Production of bioenergy and biochemicals from industrial and agricultural wastewater. *Trends Biotechnol., 22*, 477–485.

Balat, H., & Kırtay, E., (2010). Hydrogen from biomass—present scenario and future prospects. *Int. J. Hydrogen Energy, 35*, 7416–7426.

Bhatia, L., Sarangi, P. K., & Nanda, S., (2020). Current advancements in microbial fuel cell technologies. In: Nanda, S., Vo, D. V. N., & Sarangi, P. K., (eds.), *Biorefinery of Alternative Resources: Targeting Green Fuels and Platform Chemicals* (pp. 477–494). Springer Nature: Singapore.

Boopathy, R., & Daniels, L., (1991). Isolation and characterization of a furfural degrading sulfate-reducing bacterium from an anaerobic digester. *Curr. Microbiol., 23*, 327–332.

Cavinato, C., Fatone, F., Bolzonella, D., & Pavan, P., (2010). Thermophilic anaerobic codigestion of cattle manure with agro-wastes and energy crops: Comparison of pilot and full scale experiences. *Bioresour. Technol., 101*, 545–550.

Cavinato, C., Giuliano, A., Bolzonella, D., Pavan, P., & Cecchi, F., (2012). Bio-hythane production from food waste by dark fermentation coupled with anaerobic digestion process: A long-term pilot scale experience. *Int. J. Hydrogen Energy, 37*, 11549–11555.

Chen, C. C., Chuang, Y. S., Lin, C. Y., Lay, C. H., & Sen, B., (2012). Thermophilic dark fermentation of untreated rice straw using mixed cultures for hydrogen production. *Int. J. Hydrogen Energy, 37*, 15540–15546.

Datar, R. P., Shenkman, R. M., Cateni, B. G., Huhnke, R. L., & Lewis, R. S., (2004). Fermentation of biomass-generated producer gas to ethanol. *Biotechnol. Bioeng., 86*, 587–594.

Escapa, A., Mateos, R., Martínez, E. J. J., & Blanes, J., (2016). Microbial electrolysis cells: An emerging technology for wastewater treatment and energy recovery. From laboratory to pilot plant and beyond. *Renew. Sustain. Energy Rev., 55*, 942–956.

Faloye, F. D., Gueguim, K. E. B., & Schmidt, S., (2014). Optimization of biohydrogen inoculum development via a hybrid pH and microwave treatment technique-semipilot scale production assessment. *Int. J. Hydrogen Energy, 39*, 5607–5616.

Ghimire, A., Frunzo, L., Pirozzi, F., Trably, E., Escudie, R., Lens, P. N. L., & Esposito, G., (2015). A review of dark fermentative biohydrogen production from organic biomass: Process parameters and use of by-products. *Appl. Energy, 144*, 73–95.

Grimes, C. A., Varghese, O. K., & Ranjan, S., (2008). *Light, Water, Hydrogen-The Solar Generation of Hydrogen by Water Photoelectrolysis.* Springer: New York.

Guo, X. M., Trably, E., Latrille, E., Carrere, H., & Steyer, J. P., (2010). Hydrogen production from agricultural waste by dark fermentation: A review. *Int. J. Hydrogen Energy, 35*, 10660–10673.

Hallenbeck, P. C., & Ghosh, D., (2009). Advances in fermentative biohydrogen production: The way forward? *Trends Biotechnol., 27*, 287–297.

Han, H., Wei, L., Liu, B., Yang, H., & Shen, J., (2012). Optimization of biohydrogen production from soybean straw using anaerobic mixed bacteria. *Int. J. Hydrogen Energy, 37*, 13200–13208.

Hawkes, F. R., Dinsdale, R., Hawkes, D. L., & Hussy, I., (2002). Sustainable fermentative hydrogen production: Challenges for process optimization. *Int. J. Hydrogen Energy, 27*, 1339–1347.

Jariyaboon, R., Sompong, O., & Kongjan, P., (2015). Bio-hydrogen and biomethane potentials of skim latex serum in batch thermophilic two-stage anaerobic digestion. *Bioresour. Technol., 198*, 198–206.

Kongjan, P., O-Thong, S., & Angelidaki, I., (2011). Performance and microbial community analysis of two-stage process with extreme thermophilic hydrogen and thermophilic methane production from hydrolysate in UASB reactors. *Bioresour. Technol., 102*, 4028–4035.

Kongjan, P., O-Thong, S., & Angelidaki, I., (2013). Hydrogen and methane production from desugared molasses using a two-stage thermophilic anaerobic process. *Eng. Life Sci., 13*, 118–125.

Kongjan, P., Sompong, O., & Angelidaki, I., (2011). Performance and microbial community analysis of two-stage process with extreme thermophilic hydrogen and thermophilic methane production from hydrolysate in UASB reactors. *Bioresour. Technol., 102*, 4028–4035.

Kotsopoulos, T. A., Zeng, R. J., & Angelidaki, I., (2006). Biohydrogen production in granular up-flow anaerobic sludge blanket (UASB) reactors with mixed cultures under hyper-thermophilic temperature (70 degrees C). *Biotechnol. Bioeng., 94*, 296–302.

Kumar, G., Sivagurunathan, P., Sen, B., Mudhoo, A., Davila-Vazquez, G., Wang, G., & Kim, S. H., (2017). Research and development perspectives of lignocellulose-based biohydrogen production. *Int. Biodeter. Biodegr., 119*, 225–238.

Lee, D. Y., Ebie, Y., Xu, K. Q., Li, Y. Y., & Inamori, Y., (2010). Continuous H_2 and CH_4 production from high-solid food waste in the two stage thermophilic fermentation process with the recirculation of digester sludge. *Bioresour. Technol., 101*, S42–S47.

Levin, D. B., Pitt, L., & Love, M., (2004). Biohydrogen production: Prospects and limitations to practical application *Int. J. Hydrogen Energy, 29*, 173–185.

Lin, C. Y., Lay, C. H., Sen, B., Chu, C. Y., Kumar, G., Chen, C. C., Kumar, G., et al., (2012). Fermentative hydrogen production from wastewaters: A review and prognosis. *Int. J. Hydrogen Energy, 37*, 15632–15642.

Liu, Z., Si, B., Li, J., He, J., Zhang, C., Lu, Y., Zhang, Y., & Xing, X. H., (2018). Bioprocess engineering for biohythane production from low-grade waste biomass: Technical challenges towards scale-up. *Curr. Opin. Biotechnol., 50*, 25–31.

Liu, Z., Zhang, C., Lu, Y., Wu, X., Wang, L., Wang, L., Han, B., & Xing, X. H., (2013). States and challenges for high-value biohythane production from waste biomass by dark fermentation technology. *Bioresour. Technol., 135*, 292–303.

Logan, B. E., Hamelers, B., Rozendal, R., Schröder, U., Keller, J., Freguia, S., Aelterman, P., et al., (2006). Microbial fuel cells: Methodology and technology. *Environ. Sci. Technol., 40*, 5181–5192.

Łukajtis, R., Hołowacz, I., Kucharska, K., Glinka, M., Rybarczyk, P., Przyjazny, A., & Kamiński, A., (2018). Hydrogen production from biomass using dark fermentation. *Renew. Sustain. Energy Rev., 91*, 665–694.

Luo, G., Talebnia, F., Karakashev, D., Xie, L., Zhou, Q., & Angelidaki, I., (2011). Enhanced bioenergy recovery from rapeseed plant in a biorefinery concept. *Bioresour. Technol., 102*, 1433–1439.

Magnusson, L., Islam, R., Sparling, R., Levin, D., & Cicek, N., (2008). Direct hydrogen production from cellulosic waste materials with a single-step dark fermentation process. *Int. J. Hydrogen Energy, 33*, 5398–5403.

Mamimin, C., Prasertsan, P., Kongjan, P., & Sompong, O., (2017). Effects of volatile fatty acids in biohydrogen effluent on biohythane production from palm oil mill effluent under thermophilic condition. *Electr. J. Biotechnol., 29*, 78–85.

Mamimin, C., Singkhala, A., Kongjan, P., Suraraksa, B., Prasertsan, P., Imai, T., & O-Thong, S., (2015). Two stage thermophilic fermentation and mesophilic methanogen process for biohythane production from palm oil mill effluent. *Int. J. Hydrogen Energy, 40*, 6319–6328.

Moodley, P., & Kana, E. B. G., (2015). Optimization of xylose and glucose production from sugarcane leaves (*Saccharum offinarum*) using hybrid pretreatment techniques and assessment for hydrogen generation at semi-pilot scale. *Int. J. Hydrogen Energy, 40*, 3859–3867.

Najafpour, G., Younesi, H., & Mohamed, A. R., (2004). Effect of organic substrate on hydrogen practical application. *Int. J. Hydrogen Energy, 29*, 173–185.

Nanda, S., Rana, R., Vo, D. V. N., Sarangi, P. K., Nguyen, T. D., Dalai, A. K., & Kozinski, J. A., (2020). A spotlight on butanol and propanol as next-generation synthetic fuels. In: Nanda, S., Vo, D. V. N., & Sarangi, P. K., (eds.), *Biorefinery of Alternative Resources: Targeting Green Fuels and Platform Chemicals* (pp. 105–126). Singapore: Springer Nature.

Nandi, R., & Sengupta, S., (1998). Microbial production of hydrogen: An overview. *Crit. Rev. Microbiol., 24*, 61e84.

O-Thong, S., Suksong, W., Promnuan, K., Thipmunee, M., Mamimin, C., & Prasertsan, P., (2016). Two stage thermophilic fermentation and mesophilic methanogenic process for biohythane production from palm oil mill effluent with methanogenic effluent recirculation for pH control. *Int. J. Hydrogen Energy, 41*, 21702–21712.

Pasin, T. M., De Almeida, P. Z., De Almeida, S. A. S., Da Conceição, I. J., & De Teixeira, D. M. P. M. L., (2020). Bioconversion of agro-industrial residues to second-generation bioethanol. In: Nanda, S., Vo, D. V. N., & Sarangi, P. K., (eds.), *Biorefinery of Alternative Resources: Targeting Green Fuels and Platform Chemicals* (pp. 23–47). Singapore, Springer Nature.

Porpatham, E., Ramesh, A., & Nagalingam, B., (2007). Effect of hydrogen addition on the performance of a biogas fueled spark-ignition engine. *Int. J. Hydrogen Energy, 32*, 2057–2065.

Sarangi, P. K., & Nanda, S., (2018). Recent developments and challenges of acetone-butanol-ethanol fermentation. In: Sarangi, P. K., Nanda, S., & Mohanty, P., (eds.), *Recent Advancements in Biofuels and Bioenergy Utilization* (pp. 111–123). Singapore: Springer Nature.

Sarangi, P. K., & Nanda, S., (2020a). Biohydrogen production through dark fermentation. *Chem. Eng. Technol., 43*, 601–612.

Sarangi, P. K., & Nanda, S., (2020b). Bioconversion waste biomass to bioethanol. In: Sarangi, P. K., & Nanda, S., (eds.), *Bioprocessing of Biofuels* (pp. 25–33). Boca Raton: CRC Press.

Sarangi, P. K., Nanda, S., & Vo, D. V. N., (2020). Technological advancements in the production and application of biomethanol. In: Nanda, S., Vo, D. V. N., & Sarangi, P. K., (eds.), *Biorefinery of Alternative Resources: Targeting Green Fuels and Platform Chemicals* (pp. 127–140). Singapore: Springer Nature.

Si, B. C., Li, J. M., Zhu, Z. B., Zhang, Y. H., Lu, J. W., Shen, R. X., Zhang, C., Xing, X. H., & Liu, Z., (2016). Continuous production of biohythane from hydrothermal liquefied cornstalk biomass via two-stage high-rate anaerobic reactors. *Biotechnol. Biofuels, 9*, 254.

Sompong, O., Mamimin, C., & Prasertsan, P., (2018). Biohythane production from organic wastes by two-stage anaerobic fermentation technology. *Adv. Biofuels Bioenergy*, 83.

Sompong, O., Suksong, W., Promnuan, K., Thipmunee, M., Mamimin, C., & Prasertsan, P., (2016). Two-stage thermophilic fermentation and mesophilic methanogenic process for biohythane production from palm oil mill effluent with methanogenic effluent recirculation for pH control. *Int. J. Hydrogen Energy, 41*, 21702–21712.

Sørensen, B., (2005). *Hydrogen and Fuel Cells—Emerging Technologies and Applications*. New York: Elsevier Academic Press.

Suksong, W., Kongjan, P., & Sompong, O., (2015). Biohythane production from co-digestion of palm oil mill effluent with solid residues by two-stage solid-state anaerobic digestion process. *Energy Proc., 79*, 943–949.

Temudo, M. F., Kleerebezem, R., & Van, L. M., (2007). Influence of the pH on (open) mixed culture fermentation of glucose: A chemostat study. *Biotechnol. Bioeng., 98*, 69–79.

Wang, X., & Zhao, Y. C., (2009). A bench-scale study of fermentative hydrogen and methane production from food waste in integrated two-stage process. *Int. J. Hydrogen Energy, 34*, 245–254.

Willquist, K., Nkemka, V. N., Svensson, H., Pawar, S., Ljunggren, M., Karlsson, H., Murto, M., et al., (2012). Design of a novel biohythane process with high H_2 and CH_4 production rates. *Int. J. Hydrogen Energy, 37*, 17749–17762.

Wu, K. J., & Chang, J. S., (2007). Batch and continuous fermentative production of hydrogen with anaerobic sludge entrapped in a composite polymeric matrix. *Proc. Biochem., 42*, 279–284.

CHAPTER 8

Socio-Economic and Techno-Economic Aspects of Biomethane and Biohydrogen

RANJITA SWAIN,[1] RUDRA NARAYAN MOHAPATRO,[1] and BISWA R. PATRA[1,2]

[1]*Department of Chemical Engineering, C.V. Raman Global University, Bhubaneswar, Odisha, India*
E-mail: ranjitaswain79@gmail.com (Ranjita Swain)

[2]*Department of Chemical and Biological Engineering, University of Saskatchewan, Saskatoon, Saskatchewan, Canada*

ABSTRACT

Nonrenewable energy sources are depleting day by day. The production and utilization of biomethane and biohydrogen are prospective with several technical challenges. Many methodologies are adopted by the researchers to produce biomethane and biohydrogen from numerous resources. Biomethane and biohydrogen production from different sources is an alternative source of energy instead of fossil fuel. Biomethane is one of the solutions of decarburization. Global product awareness has been increased to reduce greenhouse gas emissions from automobiles and make the transportation sector environmentally friendly. In this chapter, the socio-economic and techno-economic impacts of biohydrogen and biomethane are discussed.

8.1 INTRODUCTION

Fossil fuels are being used in different areas such as in transportation, cooking, industrial manufacturing and energy production. The unprecedented and exponential increase in energy demand is depleting the fossil fuel reserves, and their burning deteriorates the environment. Therefore, to cope with the current energy demand, many researchers are exploring new sustainable alternative energy to substitute fossil fuels. The emission of greenhouse gasses from different sources creates environmental pollution and also affects social life (Nanda et al., 2016). The fuel like biomethane and biohydrogen are derived from biomass that has been helped to reduce the harmful impact on environmental and social life. It is a promising strategy to produce biomethane as a biofuel from the waste generated daily from domestic and industries. The most available raw materials such as agricultural and forest residues, municipal solid waste, food waste, animal manure and microalgae can be used for the production of biofuels (Nanda et al., 2016a, 2016b; Barbot, 2016; Shafiei, 2018; Ardolino and Arena, 2019; Prussi, 2019; Okolie et al., 2021; Nanda and Berruti, 2021). This is one of the concepts for supporting a circular economy. This is considered during the utility of biomethane in domestic use as fuel or to generate power (Molino, 2013; Morero, 2015; Morgan, 2017). The anaerobic digestion is the biological methodology for the production of biomethane from locally available waste feedstocks. This process is more sustainable and energy-efficient (Molino, 2013; Pierie, 2015; Raffiaani, 2017; Li, 2017; Florio, 2019).

The production of biofuel by using different systems can be applied to assess different fuels such as biomethane, hydrogen, methanol, dimethyl ether (DME) and hydrocarbon fuels from the Fischer-Tropsch (FT) process. These biofuels can be produced through standalone gasification and polygeneration by integrating gasification with combined heat and power plants.

Among all the alternative energy sources, biohydrogen is considered as one of the promising alternative energy for the future. Biohydrogen is the cleanest fuel with zero carbon footprint and viable fuel for automobiles and electricity generation via fuel cells. Hydrogen is also a highly used feedstock in many industries like petroleum refinery, chemical, food, and steel (Nanda et al., 2017). The primary sources for hydrogen production are natural gas, coal, oil, which share a major portion, while other sources for hydrogen production include biomass feedstock. Hydrogen is economically more favorable for production from fossil fuels but lacking in the technical aspect for production from biomass (Kumar and Shukla, 2016; Li, 2017; Kraussler, 2018; Salman, 2019). The current nonrenewable resources used for hydrogen production through reforming are industrially viable but pose environmental threats through the emissions of greenhouse gas (Singh et al., 2018). The rapid increase in global demand for hydrogen is evident that in mitigating emissions issues, hydrogen plays a distinct role. According to International Renewable Energy Agency's (IRENA) renewable energy roadmap, the total share of final global energy consumption will reach 6% by the year 2050 (IRENA 2019).

Socio-economic and techno-economic analysis and measurement for the production of biomethane can be applied to waste valorization by anaerobic digestion (Collet, 2016; Pestalozzi, 2019; Adnan, 2019). This chapter is based on socio-economic and techno-economic studies of biomethane and biohydrogen production processes.

8.2 PROCESS DEVELOPMENT FOR TECHNO-ECONOMIC ANALYSIS

8.2.1 *PROCESS FLOW*

The process flowsheet for the techno-economic feasibility study shown in Figure 8.1 indicates the plan for the setting of a biomethane plant newly or through upgrading the existing plant. It can justify that the step forwarded with the process development by planning. Then the requirements for setting up plants are to be noted down. The location is one of the priorities to establish a plant where transportation, raw material availability, labor, and other facilities should be taken into consideration. The process should be profitable, which can be known from an economic assessment (Shafiei, 2018).

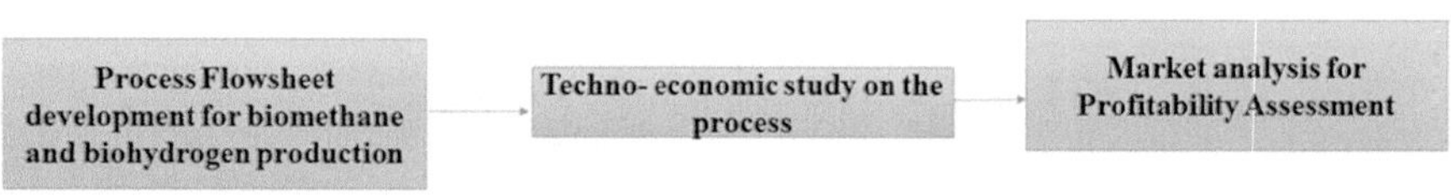

FIGURE 8.1 Flow chart for techno-economic analysis.

8.2.2 *ESTIMATION OF TOTAL CAPITAL INVESTMENT AND OPERATING COST*

Biomethane is a traditional biofuel widely used in many developing countries. It is found that upgrading an existing plant is not profitable than the planned biomethane plant in a construction mode. The profit can be measured based on the discounted cash flow (DCF) method and the net present value (NPV). However, the break-even point (BEP) can give the subsidy of biogas and biomethane by which the plant can reach its profit margin (Kraussler, 2018; Ferella, 2018). Biogas production for power generation and in rural areas, especially in Asian countries. The techno-economic analysis shows that the payback period for this system implementation is in between 2.6 and 4 years (Khan, 2014; Kraussler, 2018).

Figure 8.2 indicates that total capital investment is calculated based on fixed cost and working capital cost. The fixed cost is being calculated based on direct cost and indirect cost. The direct cost describes the expenses towards land, building, equipment cost with installation charges, control system, instrument cost with installation, piping, electrical, and other facilities. The indirect cost includes temporary construction of roads, buildings, engineering, and supervision, administration, personnel, contingencies, contractor fees and security costs. The operating cost is the factor multiplied with the total value for purchased equipment (Shafiei, 2018).

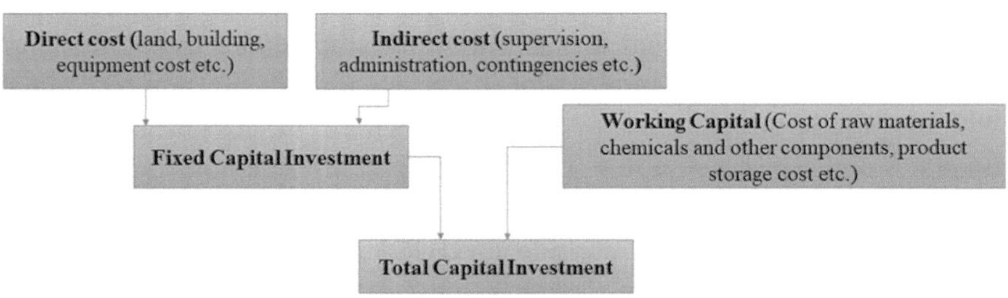

FIGURE 8.2 Flow chart for calculating total capital investment (TCI).

Working capital is a part of the project cost. It contains operational costs to produce the finished product. The working capital comprises the cost of raw materials, the supply of chemicals and other components, finishing the cost of product and storage cost of products, etc., used in the process. Working capital also includes the cash paid for taxes, accounts, and operating expenditure. Total production cost is based on total capital investment and other expenses of the project. Total production cost includes the manufacturing expenditure and the general cost. Fixed costs, direct production expenditure, and plant overhead combine to form the total manufacturing expenses. The raw materials used for the process, operation cost for labor and supervisors, maintenance, royalties, and patents. The fixed cost contains taxes due to income, rent payment if any, plant insurance, and depreciation. The general expenses could include research and development, executive salaries, marketing cost, product price, etc. (Shafiei, 2018).

8.2.3 PROFITABILITY ANALYSIS

As the project is designed and the cost estimated, the next step could be to establish the plant. However, the plant usually sets up while the profit analysis is in favor. Hence, the process is compared with other processes based on its cost estimation and profit analysis. The plant produces biomethane from different sources by using different raw materials, can be compared by their profitability study. This profitability study is mostly analyzed by using a techno-economic feasibility study for the entire plant.

The techno-economic feasibility study is analyzed by comparing the profitability of different process methodologies applied for the production of biomethane. In addition, there is another comparison study between the new plant or upgrading of existing plants to find the convenient route for the production of biomethane. The cost estimation is evaluated by using a flow diagram of all estimations required for finding the total project cost, the cost estimate for operation, and the value of money concerning time. Time-value money depends on the discount and interest rate. The production cost can be considered by using the total product cost divided by total plant capacity (Shafiei, 2018). The calculation of profit depends on different parameters such as net present value, break-even point, discounted cash flow rate of return and the payback period. Figure 8.3 shows the method to calculate total product cost (TPC).

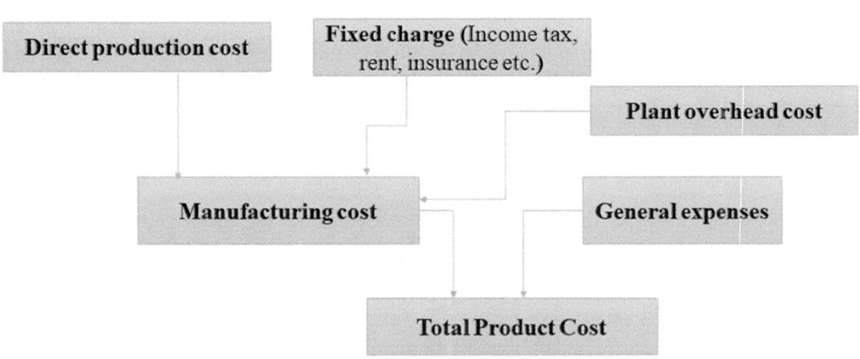

FIGURE 8.3 Flow chart developed for calculating total product cost (TPC).

Figure 8.3 is plotted between time and cost for the representation of the break-even point and payback period. The production may start at zero profit. This shows the loss is there and shown below is the baseline of the production. The break-even point pattern moves upward when the industry makes a profit from the production. At a certain point, the line may touch the baseline, which is called the break-even point. The payback period is also counted at this point from which the profit of the plant begins. Hence, it proves that the profitability depends on each factor, including the installation of the plant.

8.2.4 INVESTMENT PARAMETERS

The parameters involved in the plant set up that should be stated clearly. The techno-economic feasibility study depends on the investment parameters. The financial feasibility study may be changed by changing the constraints like the rate of taxation and interest. These two parameters are very much necessary to get the rate of return. Depreciation is one of the factors that depend on the plant set up area and the production procedures (Shafiei, 2018).

8.3 TECHNO-ECONOMIC AND SOCIO-ECONOMIC ASPECTS OF BIOMETHANE PRODUCTION

A flow chart of the biomethane production process is shown in Figure 8.4. This is the flowchart for the production biomethane by using two types of biomass that endure anaerobic digestion. Here the biomass is vegetable residue from livestock and agri-food industry. These are to be grouped into pumpable and nonpumpable feedstock. The optimal mixture of sludge for codigestion helps in improving the digestion process to yield biomethane (Adnan, 2011; Daniel, 2018; Admano, 2019). The anaerobic digestion process is one of the well-known processes to produce biomethane by using organic wastes and energy crops. This process is mature technology to generate biofuels. Nowadays, different types of upgrading technologies are taken up by the industries. The methods are using to purify the biogas to get separate the biomethane from the off-gasses. The methods for removing different gasses by using different techniques such as water scrubbing, organic physical scrubbing, amine scrubbing, pressure swing adsorption, membrane technology, etc.

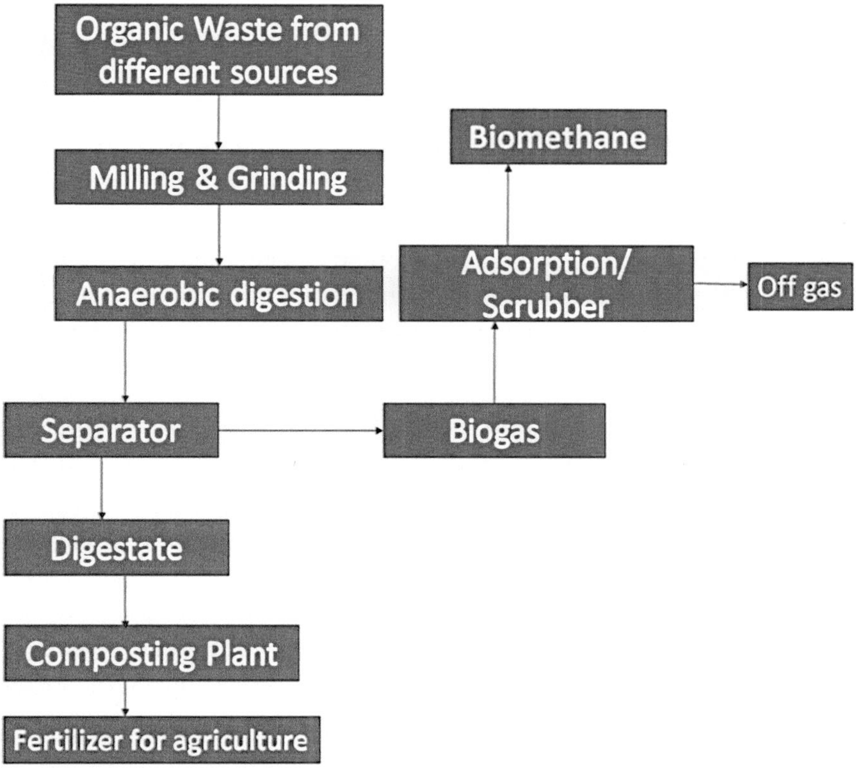

FIGURE 8.4 Production of biomethane from organic wastes.

8.3.1 *TECHNO-ECONOMIC STUDIES*

Costing of biomethane production plant consists of three factors such as: (i) biogas production, (ii) upgrading, and (iii) compression of gas and its distribution. The costing of additional plants such as the production of digestate and the fueling stations is also considered for total production cost. Manufacturing cost also includes maintenance and overhead charges. The plant gains more profit when the combined work can be done between the producer and methane distributor (Adnan, 2011; Pierei, 2015; Daniel, 2018; Admano, 2019). Methane is recovered and purified from this biogas plant by using upgrading methods. Three methods are used for producing biomethane, such as pressure water scrubbing, hot potassium carbonate and pressure swing adsorption (Adnan, 2011; Daniel, 2018; Barbera, 2018).

Most of the plants in the European country have their studies on two possibilities to calculate the profitability by using a techno-economic investigation. This study on two cases include: (i) upgrading the existing biogas plant to biomethane, and (ii) new plant for producing biomethane from organic fraction of municipal solid waste (OFMSW) as substrate (Kraussler, 2018; Ferella, 2019). It is found that the new biomethane plant was set up with a capacity of 250 m^3/h, where the major substrate was OFMSW. This newly established biomethane plant was profitable than biomethane production by upgrading the existing biogas plant. The existing biogas plants can be used to generate green electricity, which also increases the part of renewable sources in the power division. The new pants contribute their share in conveyance. In every case, energy production is a part of internal utilities (Kraussler, 2018; Ferella, 2019). The biogas upgrade to producing biomethane for the betterment of the society in Argentina and techno-economic study on the up-gradation of the process. Techno-economic investigation shows that the upgrading of biomethane from biogas is more profitable by using the water as the solvent. This is more cost-effective than other upgrading processes using diglycolamine (DGA) and dimethyl ethers of polyethelyne glycol (DEPG) as solvents.

There is a fluctuating electricity price of a novel operation for an existing biogas plant. In the biogas plant, the excess electricity consumed from an electrolyzer during the production of renewable energy can be stored as methane by methanation of hydrogen with CO_2. The important components are the electrolyzer, electric grid, an anaerobic digestor, and the biogas upgrading unit. It is also seen that the economic value increases due to the electricity consumption in upgrading the existing plant for biomethane production. The electrolyzer can be used for the production of methane by the methanation process, and the cost of the electrolyzer investment can be reduced to 60–72% by using the same electricity with an increase in methane price from 20–76% (Li, 2017; Florio, 2019; Pääkkönena, 2017).

8.3.2 SOCIO-ECONOMIC STUDIES

Social benefits are being countable whenever the product is distributed among the consumers. Biomethane storage is one of the major parts of the plant production system. Storage is also a great challenge for the biomethane production plants as it is costly. This system can be an efficient one to serve the society in different utility areas like power, heat, transport, and industry. The storage methods are flexible with different prices and efficiencies. The economy may be increased for this storage system. To overcome from storage cost, the storage of biomethane is to be combined with the grid service (Budzianowski, 2017; Ardolino and Arena, 2019). The storage system is connected to the plant process flowsheet to optimize the developed technology to reduce the cost. Biomethane storage systems are viable to reduce the environmental impacts, increases the economic viability, and also bring social benefits (Budzianowski, 2017). The development of biomethane plants has an impact on many aspects such as environmental, economic, and social. The production and utilization of biomethane that gives benefits to the farmers and local communities. Social security is the security that society furnishes through an appropriate organization.

Biomethane is a renewable source of fuel that can replace the fossil fuel, used in the energy and transport sectors and contribute to the supply of energy. It protects and improves the natural resources and environment of the local area. The local community can use this fuel for their daily basic needs within their limitations so that they are less dependent on other countries or areas. This can reduce the emission of CO_2 and other off-gasses, which diminishes global warming. The organic wastes from different sources get treated to produce the biogas to biomethane. The byproduct generated during the process can be used as a fertilizer. This technology can help in reducing waste and also diminish the cost required for their disposal. The energy from renewable sources minimizes green-house gasses emission and sustainable waste management. Wastewater treatment plants also decrease the water pollution in soil and ground. The biomethane plant set up in any area needs staff for different purposes like for collection and transportation of raw materials, equipment handling, manufacturing, operation, maintenance, and construction, etc. The biomethane production can be more favorable for farmers to remediate the agricultural crop residues and cattle manure while producing biofuels to supplement their basic household needs such as lightening, cooking, and heating. This can give additional economic growth for a rural community. The plant requires more manpower than decreases the unemployment level in the country.

The up-gradation of biogas plants can create job opportunities and solves the problems for the biological waste generated from different sources like industries, farms, municipalities, etc. The biomethane production helps in plant installation, which can offer workplaces for the local community. This raises the income and motivation to get employed within the state. The income amount and workplace specification depend on the production capacity of the plant and also the automation level. The industrial set up and the workplace motivate the next generation to enhance their education and also to gain practical knowledge and develop their skills. The storage and distribution of biomethane need a fuel filling station where it is stored in a compressed form. Hence, the stations should be improved and maintained, which in both cases generate the employment and welfare facility in the state.

The digestate from biomethane production plant can be used in different areas in the form of pellets to produce energy, as an organic fertilizer for crop production and used as building materials. The organic fertilizer is used in local areas reduces the nitrates in water bodies. The organic fuels (pellets) used for energy production decreases the amount of CO_2 in the atmosphere. The digestate can be used as a filler material (as dry fibers) for plywood works in buildings. In this way, the digestate is one of the resources of the economy for the local people (Paolini, 2018; Karkalina, 2019).

The biogas upgrading is to produce biomethane for the betterment of the society in Argentina and techno-economic study on the up-gradation of the process. The biomethane production using the up-gradation of biogas using the different adsorbents such as water, dimethyl ethers of polyethelyne glycol and diglycolamine solvents. The amine process is avoidable due to its impact on climate change and human toxicity. This process generates ethylene oxide that affects society. In the water process, the impact is less than other solvents. This slight harmful effect is due to losses of methane during the biogas up-gradation method (Morero, 2015). The anaerobic digestion process to produce biomethane from different biomass and wastes is one of the sustainable and renewable energies (Pierie, 2015; Rotunn, 2016).

Society awareness is most important for the utilization of biomethane in different sectors. The total production cost of biomethane depends upon operational cost, raw materials, and other costs. There are some factors which influence the income of the biomethane producers and increasing the economic conditions of the plant. Storage and transportation costs affect the economics of the plant. It can be reduced by distributing by using the pipeline grid of natural gas if the pipelines are available nearby. The government policies and certification are also one factor for the utility of biomethane by the consumers. The global political initiative can help the producers make a profit in the long run as well as in changing the climate for zero carbon footprints.

8.4 TECHNO-ECONOMIC AND SOCIO-ECONOMIC ASPECTS OF BIOHYDROGEN PRODUCTION

The production of biohydrogen is trending in the current scenario because of its source, which is from renewable and sustainable materials. Figure 8.5 illustrates hydrogen production technologies. Biohydrogen production involves different biological and thermochemical conversion pathways. Biohydrogen from biomass via different pathways is found to be promising. The first category of biomass i.e., energy crops, are specially cultivated for energy purposes (Nanda et al., 2013). The second is forestry waste consisting of dead trees, branches, leaves, roots, etc. The third category is agricultural waste consisting of food crops and animal waste, and the fourth one is municipal solid waste and industrial wastes. Thermochemical conversion technology, like pyrolysis and gasification, are used to convert biomass to bioenergy. Such processes generate gaseous products, which undergo the reforming process to maximize the production of hydrogen.

In the electrochemical hydrogen production process, water is split into hydrogen and oxygen using an electric current. The resulting hydrogen gas via electrolysis is considered as renewable if the electricity used for the process is produced via renewable sources like solar, hydrothermal, wind. The thermochemical conversion techniques (e.g., pyrolysis and gasification) convert a wide range of biomass into bioenergy (solid, liquid, and gas fuels) using heat. Thermochemical processes are classified based on operational conditions such as process temperature, pressure, reaction time, oxidizing agent, and desired products. The hydrogen production using liquid fuels like ethanol are considered renewable. The liquid fuels are reacted with steam at high temperatures to produce biohydrogen. Water is split via chemical reactions at elevated temperatures generated by nuclear reactors or solar concentrators to produce hydrogen. Photoelectrochemical systems can also be used to split water for the production of biohydrogen using solar energy or semiconductors.

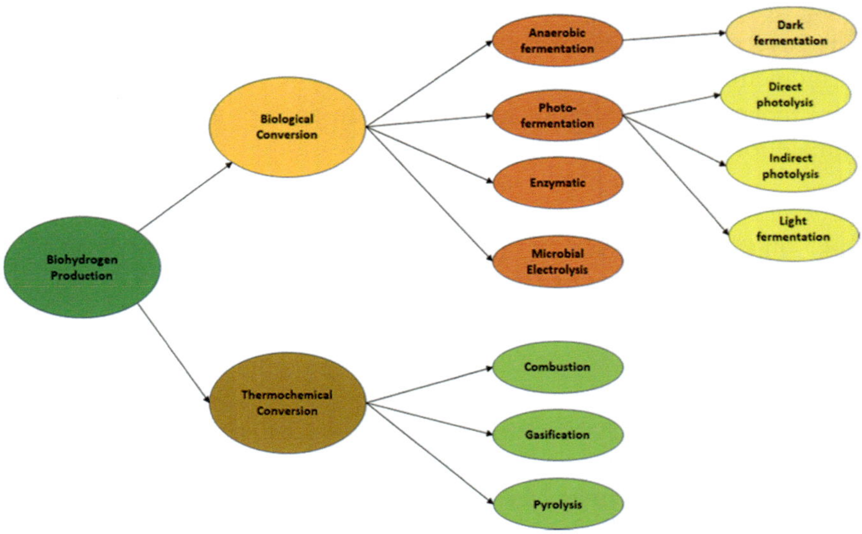

FIGURE 8.5 Classification of biohydrogen production pathways.

The biohydrogen production using biological pathways has seen increasing significantly. Dark fermentation and photo-fermentation using microorganisms like algae and bacteria are used for the production of low-cost biohydrogen from organic substrates (Sarangi and Nanda, 2020). The fermentation process for biohydrogen production seems to be more promising and favorable because of the eco-friendly conversion process and high yield product. The production of low-cost biohydrogen is possible by using organic wastes and carbohydrates augmented wastewater as substrates. Over the decades, extensive studies in the field of dark fermentation showed effective results in biohydrogen production.

Biomass precursors are generally converted into sugar or carbohydrate and then fermented by anaerobic organisms for the production of biohydrogen. The bacterial species such as *Clostridium pasteurianum*, *Clostridium beijerinckii* and *Clostridium acetobutylicum* are the key microorganisms responsible for biohydrogen production (Cabrol et al., 2017). Photo-heterotrophic bacteria convert organic acids into hydrogen in the existence of light (Sarangi and Nanda, 2020). Photosynthetic bacteria like *Rhodobacter sphaeroides*, *Rhodospirillum rubrum* and *Rhodopseudomonas palustris* are also found to be efficient microorganisms for the production of biohydrogen (Kapdan and Kargi, 2006). Photo-bioreactors like bubble column and tubular reactors are often used for the production of biohydrogen.

The photosynthesis process is used to split water molecules into oxygen and hydrogen ions in the presence of algae. Furthermore, the hydrogenase enzyme converts hydrogen ions into hydrogen gas. There are different types of algae, such as *Chlamydomonas reinhardtii* and green algae (e.g., *Scenedesmus obliquus*, *Platymonas subcordiformis*, etc.), involved in this process. The production of biohydrogen from biomass compounds like cellulose and sucrose takes place by adding appropriate diverse enzymes Microbial electrolysis cell (MEC) is an emerging method for hydrogen production. In MEC, electrogenic bacteria oxidize wastewater and organic matter to generate protons and electrons. Electrodes are used to produce hydrogen upon passing electric current.

8.4.1 TECHNO-ECONOMIC STUDIES

Techno-economic analysis for biohydrogen production pathways is essential for its commercialization. Bartels et al. (2010) evaluated the economic impact of hydrogen generated from conventional fossil fuel and different alternative sources. The study showed that coal and natural gas led hydrogen production as most effective owing to its economic aspect. The cost of hydrogen generated from coal was $0.36–1.8/kg, whereas it was $2.4–3.4/kg from natural gas. The most expensive source for hydrogen production was produced using solar and wind energy because of its high capital investment. Many researchers have shown biomass feedstocks as the cheapest source for the production of hydrogen (Amigun et al., 2010; Han et al., 2016; Offer et al., 2011; Sarkar and Kumar, 2010). However, the production cost of hydrogen is dependent on many factors like plant capacity and location, feedstock price and availability, marketability, and transportation. The rising environmental and sustainability concern because of fossil fuels has shifted the focus towards the production of renewable biohydrogen from biomass, which could address the fuel demand of the automobile market in the future. Table 8.1 summarizes the pathways for the production of biohydrogen with plant capacity and production cost.

TABLE 8.1 Biohydrogen Production Pathways from Different Sources with Its Capacity and Production Cost

Production pathways	Source of energy	Production capacity of hydrogen (tons/day)	Estimated hydrogen cost	References
Dark fermentation	Biogas	0.09	\$6.70/kg	Abusoglu et al. (2016)
High-temperature steam electrolyzer	Biogas	0.8	\$2.40/kg	
Dark fermentation (combined with solid-state fermentation)	Biomass	10	\$2.29/m^3	Han et al. (2016)
Enzymatic hydrolysis	Biogas	2	\$14.89/kg	Han et al. (2016)
Steam reforming	Bio-oil	2000	\$2.40/kg	Sarkar and Kumar (2010)
Gasification	Biomass	500	\$15.61/GJ	Patel et al. (2016)
Gasification	Biomass	1019.71	\$34.24/GJ	
Pyrolysis	Biomass	500	\$26.01/GJ	
Gasification	Biomass	10	\$4.60–7.80/kg	Dowaki et al. (2007)
High-temperature steam electrolyzer	Geothermal	86.4	\$1.90	Kanoglu et al. (2008)

Redwood et al. (2009) studied the mechanism and integration strategies of light and dark fermentation for the production of biohydrogen. The study reported promising results by combining different microorganisms to limit the weakness of individual species. The study was based on the dual system concept in which the initial stage was light-driven with cyanobacteria and algae. The second stage was with dark fermentation. This study showed that the use of pure culture was ineffective. In contrast, the mixed culture led dark fermentation to possess better efficiency as it eliminates the need for pretreatment of the feedstocks. It was evident from the study that the dual system enhanced low-cost biohydrogen production as compared to the single system.

Sydney et al. (2014) reported extensive research on the economic process for the production of biohydrogen via fermentation of sugarcane. The percentage yield of biohydrogen was found to be 32.7%. Han et al. (2016) evaluated techno-economic analysis for biohydrogen production using biomass (food waste) as a precursor. The production involved the dark fermentation combined with solid-state fermentation. The plant with a capacity of 10 tons/day was designed using the Aspen Plus software, and the lifetime of the plant was ten years. The production cost for the biohydrogen was found to be \$2.29/m^3 with five years of payback period and 20.2% as internal rate return. The biohydrogen cost was found to be much cheaper than the market price of \$2.7/m^3. The researchers reported that the combined process was a promising pathway for low-cost biohydrogen production, which could be scaled up for industrial processing.

Amigun et al. (2010) studied the concept of adsorption enhanced reforming for the gasification process. The study focused on the production of hydrogen-rich syngas and methanol production. The researchers studied the gasification of maize to produce hydrogen at the cost of \$34.24/GJ (plant capacity of 1020 tons/day). (Salkuyeh et al., 2018) evaluated Techno-economic analysis and life cycle analysis of hydrogen production via gasification of biomass. The researchers reported different gasification processes for the production of hydrogen and found \$0.7–0.33/kg as the lowest cost for the hydrogen production in a fluid bed gasifier. Dowaki et al. (2007) investigated the economic analysis of biohydrogen production using the gasification method. The investigators reported that at a plant capacity of 10 t/d the cost of hydrogen was \$4.6–7.8/kg. Thermochemical conversion technologies produce bio-oil, which can be a prominent feedstock for biohydrogen production.

Sarkar and Kumar (2010) studied the potential application of fast pyrolysis derived bio-oil in producing pure biohydrogen. The researcher advanced a techno-economic analysis model for biohydrogen cost estimation. The bio-oil derived from different feedstocks like whole-tree biomass, agricultural waste (barley and wheat straw), and forest wastes underwent steam reforming for the biohydrogen production at 2000 tons/day as plant capacity. The production cost of hydrogen for the aforementioned biomasses was \$2.40/kg, \$4.55/kg and \$3.04, respectively. The study stated that transportation cost and capital cost for bio-oil production adds more than 50% to the biohydrogen production cost. However, considering carbon credits of CO_2 equivalent gives an upper hand to the biohydrogen to compete with the natural gas cost. Many studies carried out on bio-oil reforming using different catalysts to enhance hydrogen production (Basagiannis and Verykios, 2007; Davidian et al., 2008; Domine et al., 2008).

The biohydrogen production from bio-oil is dependent on the biomass feedstock characteristics. Carrasco et al. (2017) studied biohydrogen production from forest residue derived bio-oil. The study involved catalytic hydrotreating of bio-oil to produce hydrocarbon. The resulting pyrolysis char was subjected to the gasification process to produce syngas. For the production of biohydrogen, syngas undergoes water shifting reaction and steam reforming. The study used Aspen Plus simulation software to estimate the techno-economic aspect of a biofuel plant with a capacity of 2000 MT/day. The capital investment was \$427 million, and the fuel selling price was estimated to be \$6.25/gallon over 30 years of operation. Agricultural waste has high ash contains, and that leads to an increase in the production cost of the biohydrogen. Woody biomass and energy crops are more preferred feedstocks for low-cost biohydrogen production because of their high energy intensity.

Many studies mentioned that the additional cost for biohydrogen was the result of the high transportation cost of feedstocks to the production plant. The exploration of alternative ways to reduce the transportation cost, which takes place through trucks and trains, could bring a solution to get low-cost biohydrogen. Pipeline transportation facility for the large scale and distant plant could be a possible alternative to the costly transportation system. The scope of pipeline transportation was studied earlier, and it was found effective in reducing biohydrogen cost (Pootakham and Kumar, 2010). The integration of bio-oil production plants and biohydrogen production plants could reduce transportation costs significantly.

However, a detailed feasibility study is needed to pave a vibrant pathway for such probable solutions to produce low-cost biohydrogen in the future.

8.4.2 SOCIO-ECONOMIC STUDIES

The unprecedented increase in energy demand has depleted fossil fuel reserves. The growing environmental issues are now compelling people to go towards renewable and sustainable forms of energy. However, the alternate energy needed to be affordable for general use. Among all the alternative energies, hydrogen is considered as an efficient fuel for transportation and domestic purpose. Currently, the majority of hydrogen production comes from fossil fuels like coal and natural gas. Therefore, to mitigate environmental challenges and address global warming concerns, the focus needs to be shifted towards biohydrogen production from renewable sources like biomass, water, solar, and wind.

In a socio-economic impact perspective, biohydrogen from solar and wind is quite expensive and not affordable for use. Biohydrogen production via biological and thermochemical conversion of biomass is one of the most promising options. The commercialization of affordable biohydrogen needs lots of effort, which includes constructive planning, collaborative actions from industries, and policymakers. The success of advanced technologies is solely dependent upon public acceptance and societal awareness. Biohydrogen needs to sustain in the market amid various social and economic barriers. The consumer attitude and preference towards the utilization of biohydrogen by eliminating fossil fuel use can bring a new era. The era of green energy with zero carbon footprint.

8.5 CONCLUSIONS

Biohydrogen and biomethane are the potential resources of fuel that can be used in many sectors like transport, domestic, and industrial use. They are renewable energy like solar and wind energy. In different countries, the production of both fuels can be carried out by using different technologies. In some countries, these fuels can be produced by upgrading the existing biogas plant. Other organic wastes, agro products, algae, waste generated from industry, households, and municipalities are also potential sources for the production of biomethane. The existing plants can also be upgraded to recover biomethane from biogas by using different methods.

The consumer aspect and preference for the utilization of biomethane by eliminating fossil fuel use can bring a new era. The era of green energy with zero carbon footprint. The government policies supporting biofuels should provide the tax benefits, incentives, and investment subsidies to biomethane and biohydrogen producers or consumers, which can enhance the demands for biofuels. The global political initiative can help the producers to make a profit in the long run as a goal to the climatic change mitigation perspectives.

KEYWORDS

- **diglycolamine**
- **dimethyl ether**
- **dimethyl ethers of polyethelyne glycol**
- **Fischer-Tropsch process**
- **microbial electrolysis cell**
- **socio-economic analysis**
- **techno-economic analysis**

REFERENCES

Abuşoğlu, A., Demir, S., & Ö zahi, E., (2016). Energy and economic analyses of models developed for sustainable hydrogen production from biogas-based electricity and sewage sludge. *Int. J. Hydrogen Energy, 41*, 13426–13435.

Amigun, B., Gorgens, J., & Knoetze, H., (2010). Biomethanol production from gasification of nonwoody plant in South Africa. Optimum scale and economic performance. *Energy Policy, 38*, 312–322.

Ardolino, F., & Arena, U., (2019). Biowaste-to-biomethane: An LCA study on biogas and syngas roads. *Waste Manag., 87*, 441–453.

Barbera, E., Menegon, S., Banzato, D., D'Alpaos, C., & Bertucco, A., (2019). From biogas to biomethane: A process simulation-based techno-economic comparison of different upgrading technologies in the Italian context. *Renew. Energy, 135*, 663–673.

Barbot, Y. N., Al-Ghaili, H., & Benz, R., (2016). A review on the valorization of macroalgal wastes for biomethane production. *Mar. Drugs., 14*, 120.

Bartels, J. R., Pate, M. B., & Olson, N. K., (2010). An economic survey of hydrogen production from conventional and alternative energy sources. *Int. J. Hydrogen Energy, 35*, 8371–8384.

Basagiannis, A. C., & Verykios, X. E., (2007). Steam reforming of the aqueous fraction of bio-oil over structured Ru/MgO/Al_2O_3 catalysts. *Catal. Today, 127*, 256–264.

Budzianowski, W. M., & Brodacka, M., (2017). Biomethane storage: Evaluation of technologies, end uses, business models, and sustainability. *Energ. Convers. Manage., 141*, 254–273.

Cabrol, L., Marone, A., Tapia-Venegas, E., Steyer, J. P., Ruiz-Filippi, G., & Trably, E., (2017). Microbial ecology of fermentative hydrogen producing bioprocesses: Useful insights for driving the ecosystem function. *FEMS Microbiol. Rev., 41*, 158–181.

Carrasco, J. L., Gunukula, S., Boateng, A. A., Mullen, C. A., DeSisto, W. J., & Wheeler, M. C., (2017). Pyrolysis of forest residues: An approach to techno-economics for bio-fuel production. *Fuel, 193*, 477–484.

Collet, P., Flottes, E., Favre, A., Raynal, L., Pierre, H., Capela, S., & Peregrina, C., (2017). Techno-economic and life cycle assessment of methane production via biogas upgrading and power to gas technology. *Appl. Energy, 192*, 282–295.

D'Adamo, I., Falcone, P. M., & Ferella, F., (2019). A socio-economic analysis of biomethane in the transport sector: The case of Italy. *Waste Manag., 95*, 102–115.

Daniel-Gromke, J., Rensberg, N., Denysenko, V., Stinner, W., Schmalfuß, T., Scheftelowitz, M., & Liebetrau, J., (2018). Current developments in production and utilization of biogas and biomethane in Germany. *Chem. Ing. Tech., 90*, 17–35.

Davidian, T., Guilhaume, N., Provendier, H., & Mirodatos, C., (2008). Continuous hydrogen production by sequential catalytic cracking of acetic acid: Part II. Mechanistic features and characterization of catalysts under redox cycling. *Appl. Catal. A: Gen., 337*, 111–120.

Domine, M. E., Iojoiu, E. E., Davidian, T., Guilhaume, N., & Mirodatos, C., (2008). Hydrogen production from biomass-derived oil over monolithic Pt-and Rh-based catalysts using steam reforming and sequential cracking processes. *Catal. Today, 133*, 565–573.

Dowaki, K., Ohta, T., Kasahara, Y., Kameyama, M., Sakawaki, K., & Mori, S., (2007). An economic and energy analysis on bio-hydrogen fuel using a gasification process. *Renew. Energy, 32*, 80–94.

Ferella, F., Cucchiella, F., D'Adamo, I., & Gallucci, K., (2019). A techno-economic assessment of biogas upgrading in a developed market. *J. Clean. Prod., 210*, 945–957.

Florio, C., Fiorentino, G., Corcelli, F., Ulgiati, S., Dumontet, S., Güsewell, J., & Eltrop, L., (2019). A life cycle assessment of biomethane production from waste feedstock through different upgrading technologies. *Energies, 12*, 718.

Han, W., Hu, Y. Y., Li, S. Y., Li, F. F., & Tang, J. H., (2016). Biohydrogen production from waste bread in a continuous stirred tank reactor: A techno-economic analysis. *Bioresour. Technol., 221*, 318–323.

Han, W., Yan, Y., Gu, J., Shi, Y., Tang, J., & Li, Y., (2016). Techno-economic analysis of a novel bioprocess combining solid-state fermentation and dark fermentation for H_2 production from food waste. *Int. J. Hydrogen Energy, 41*, 22619–22625.

IRENA, (2019). *Hydrogen: A Renewable Energy Perspective*. https://www.irena.org/publications/2019/Sep/Hydrogen-A-renewable-energy-perspective (accessed on 25 June 2021).

Kapdan, I. K., & Kargi, F., (2006). Bio-hydrogen production from waste materials. *Enzyme Microb. Tech., 38*, 569–582.

Karklina, K., Slisane, D., Romagnoli, F., & Blumberga, D., (2015). Social life cycle assessment of biomethane production and distribution in Latvia. In: *Environment Technology Res. Proc. Int. Sci. Pract. Conf., 2*, 128.

Khan, E. U., Mainali, B., Martin, A., & Silveira, S., (2014). Techno-economic analysis of small scale biogas based polygeneration systems: Bangladesh case study. *Sustain. Energy Technol. Assess, 7*, 68–78.

Kraussler, M., Pontzen, F., Müller-Hagedorn, M., Nenning, L., Luisser, M., & Hofbauer, H., (2018). Techno-economic assessment of biomass-based natural gas substitutes against the background of the EU 2018 renewable energy directive. *Biomass Convers. Bioref., 8*, 935–944.

Kumar, S., & Shukla, S. K., (2016). A review on recent gasification methods for biomethane gas production. *Int. J. Energy Eng., 6*, 32–43.

Li, H., Mehmood, D., Thorin, E., & Yu, Z., (2017). Biomethane production via anaerobic digestion and biomass gasification. *Energy Proc., 105*, 1172–1177.

Molino, A., Nanna, F., Ding, Y., Bikson, B., & Braccio, G., (2013). Biomethane production by anaerobic digestion of organic waste. *Fuel, 103*, 1003–1009.

Morero, B., Groppelli, E., & Campanella, E. A., (2015). Life cycle assessment of biomethane use in Argentina. *Bioresour. Technol., 182*, 208–216.

Morgan, Jr. H. M., Xie, W., Liang, J., Mao, H., Lei, H., Ruan, R., & Bu, Q., (2018). A techno-economic evaluation of anaerobic biogas producing systems in developing countries. *Bioresour. Technol., 250*, 910–921.

Nanda, S., & Berruti, F., (2021). A technical review of bioenergy and resource recovery from municipal solid waste. *J. Hazard. Mater., 403*, 123970.

Nanda, S., Dalai, A. K., Gökalp, I., & Kozinski, J. A., (2016a). Valorization of horse manure through catalytic supercritical water gasification. *Waste Manag., 52*, 147–158.

Nanda, S., Isen, J., Dalai, A. K., & Kozinski, J. A., (2016b). Gasification of fruit wastes and agro-food residues in supercritical water. *Energy Convers. Manag., 110*, 296–306.

Nanda, S., Mohanty, P., Pant, K. K., Naik, S., Kozinski, J. A., & Dalai, A. K., (2013). Characterization of North American lignocellulosic biomass and biochars in terms of their candidacy for alternate renewable fuels. *Bioenergy Res., 6*, 663–677.

Nanda, S., Rana, R., Zheng, Y., Kozinski, J. A., & Dalai, A. K., (2017). Insights on pathways for hydrogen generation from ethanol. *Sustain. Energy Fuels, 1*, 1232–1245.

Nanda, S., Reddy, S. N., Mitra, S. K., & Kozinski, J. A., (2016). The progressive routes for carbon capture and sequestration. *Energy Sci. Eng., 4*, 99–122.

Offer, G. J., Contestabile, M., Howey, D. A., Clague, R., & Brandon, N. P., (2011). Techno-economic and behavioral analysis of battery-electric, hydrogen fuel cell and hybrid vehicles in a future sustainable road transport system in the UK. *Energy Policy, 39*, 1939–1950.

Okolie, J. A., Nanda, S., Dalai, A. K., & Kozinski, J. A., (2021). Chemistry and specialty industrial applications of lignocellulosic biomass. *Waste Biomass Valor.*, 12, 2145–2169.

Okolie, J. A., Nanda, S., Dalai, A. K., Berruti, F., & Kozinski, J. A., (2020). A review on subcritical and supercritical water gasification of biogenic, polymeric and petroleum wastes to hydrogen-rich synthesis gas. *Renew. Sust. Energy Rev., 119*, 109546.

Pääkkönen, A., Tolvanen, H., & Rintala, J., (2017). Techno-economic analysis of a power to biogas system operated based on fluctuating electricity price *Renew. Energy, 117*, 166–174.

Paolini, V., Petracchini, F., Segreto, M., Tomassetti, L., Naja, N., & Cecinato, A., (2018). Environmental impact of biogas: A short review of current knowledge. *J. Environ. Sci. Health. A, 53*, 899–906.

Patel, M., Zhang, X., & Kumar, A., (2016). Techno-economic and life cycle assessment on lignocellulosic biomass thermochemical conversion technologies: A review. *Renew. Sust. Energy Rev., 53*, 1486–1499.

Pestalozzi, J., Bieling, C., Scheer, D., & Kropp, C., (2019). Integrating power-to-gas in the biogas value chain: Analysis of stakeholder perception and risk governance requirements. *Energy Sustain. Soc., 9*, 1–18.

Pierie, F., Someren, C. E. J. V., Benders, R. M. J., Bekkeringa, J., Gemert, W. J. T. V., & Moll, H. C., (2015). Environmental and energy system analysis of bio-methane production pathways: A comparison between feedstocks and process optimizations. *Appl. Energy, 160*, 456–466.

Pootakham, T., & Kumar, A., (2010). A comparison of pipeline versus truck transport of bio-oil. *Bioresour. Technol., 101*, 414–421.

Prussi, M., Padella, M., Conton, M., Postma, E. D., & Lonza, L., (2019). Review of technologies for biomethane production and assessment of EU transport share in 2030. *J. Clean. Prod., 222*, 565–572.

Rafiaani, P., Kuppens, T., Van, D. M., Azadi, H., Lebailly, P., & Van, P. S., (2018). Social sustainability assessments in the biobased economy: Towards a systemic approach. *Renew Sust. Energ. Rev., 82*, 1839–1853.

Redwood, M. D., Paterson-Beedle, M., & Macaskie, L. E., (2009). Integrating dark and light bio-hydrogen production strategies: Towards the hydrogen economy. *Rev. Environ. Sci. Biotechnol., 8*, 149.

Rotunno, P., Lanzini, A., & Leone, P., (2016). Energy and economic analysis of water-scrubbed biogas upgrading to biomethane for grid injection and transportation application. *Renew. Energy, 102*, 417–432.

Ryckebosch, E., Drouillon, M., & Vervaeren, H., (2011). Techniques for transformation of biogas to biomethane. *Biomass Bioenergy, 35*, 1633–1645.

Salkuyeh, Y. K., Saville, B. A., & MacLean, H. L., (2018). Techno-economic analysis and life cycle assessment of hydrogen production from different biomass gasification processes. *Int. J. Hydrogen Energy, 43*, 9514–9528.

Salman, C. A., Naqvi, M., Thorin, E., & Yan, J., (2017). Impact of retrofitting existing combined heat and power plant with polygeneration of biomethane: A comparative techno-economic analysis of integrating different gasifiers. *Energy Convers. Manag., 152*, 250–265.

Sarangi, P. K., & Nanda, S., (2020). Biohydrogen production through dark fermentation. *Chem. Eng. Technol., 43*, 601–612.

Sarkar, S., & Kumar, A., (2010). Large-scale biohydrogen production from bio-oil. *Bioresour. Technol., 101*, 7350–7361.

Singh, S., Kumar, R., Setiabudi, H. D., Nanda, S., & Vo, D. V. N., (2018). Advanced synthesis strategies of mesoporous SBA-15 supported catalysts for catalytic reforming applications: A state-of-the-art review. *Appl. Catal. A: Gen., 559*, 57–74.

Sydney, E. B., Larroche, C., Novak, A. C., Nouaille, R., Sarma, S. J., Brar, S. K., & Soccol, C. R., (2014). Economic process to produce biohydrogen and volatile fatty acids by a mixed culture using vinasse from sugarcane ethanol industry as nutrient source. *Bioresour. Technol., 159*, 380–386.

Index